D0710338

WINGS OVER THE MEXICAN BORDER

University of Texas Press, Austin

WINGS OVER THE MEXICAN BORDER

Pioneer Military Aviation in the Big Bend

Kenneth Baxter Ragsdale

W
UG
634
T4
R34
1984

LC

Copyright © 1984 by the University of Texas Press
All rights reserved
Printed in the United States of America
First Edition, 1984

Requests for permission to reproduce material from this work
should be sent to Permissions, University of Texas Press, Box 7819,
Austin, Texas 78713

LIBRARY OF CONGRESS CATALOGING IN PUBLICATION DATA

Ragsdale, Kenneth Baxter, 1917–
  Wings over the Mexican border.

  Bibliography: p.
  Includes index.
  1. Aeronautics, Military—Texas—Big Bend Region.
2. United States.   Air Force—History.   3. Aeronautics,
Military—Mexico.   4. Mexico.   Fuerza Aérea Mexicana—
History.   5. Johnson, Elmo.   6. Smithers, W. D. (Wilfred
Dudley), 1895–   .   7. Big Bend Region (Tex.)—History,
Military.   8. Mexico—Frontier troubles, 1910–
I. Title.
UG634.T4R34   1984   358.4'009764'93   84-7313
ISBN 0-292-79025-2

5-2-86

Another One for Janet

Contents

". . . a one-of-a-kind story. There was just nothing else quite like Johnson's Ranch."

—Col. Howard E. Rinehart (USAF, Ret.)

PREFACE

When Col. Charles A. Lindbergh lifted his little Ryan monoplane off the rain-soaked turf of Roosevelt Field on the morning of May 20, 1927, he had no way of knowing that when he reached Paris some thirty-three hours later his achievement would alter the course of contemporary history. The world would never be the same.

Overnight Lindbergh became a world celebrity. Through an unprecedented act of bravery and courage, a hero's image had been cast in the human mind. For almost two decades, wherever men flew airplanes the helmet, goggles, and flowing white scarf would symbolize that image for an admiring public.

During the fourteen years that separated the Lindbergh flight and United States entry into World War II, the growth of aviation forms a major theme in American culture. Private, commercial, and military application of the airplane quickly captured the nation's interest, rivaling the railroad, the radio, and motion pictures in universal appeal. The impact was vast; every community, however remote, relished the sight and sound of this twentieth-century phenomenon.

It is not unusual, therefore, that the following narrative focuses on one of the most remote, rugged, and unsettled regions of the United States—the Southwest Borderlands. Into this primitive setting, its history scarred by revolution and border banditry, came modern-day military aircraft, harbingers of peace and security. Cast against a turbulent background of rebellion, free-wheeling aerial dramatics, and World War II, the story provides an intimate look into international relationships, the development of the United States aerial strike force, the character, ideals, and expectations of the men who one day would lead the nation's forces into battle, and the high esteem in which a fascinated nation held the daring young men in their flying machines.

And above the roar of airplane engines is sounded the distant rumblings of a future in which aviation would serve humanity in both peace and in a destructiveness unmatched in the history of mankind.

The focal point of the story is a Big Bend military airfield located on the Elmo Johnson ranch. Aside from the airfield itself, one of the more distinctive aspects of the study is Elmo Johnson himself, Big Bend rancher, trader, and rural sage. A civilian, he emerges as the dominant figure at one of the Air Corps' most unique facilities. Johnson sought neither rank nor authority, but ultimately filled a rare niche in the military hierarchy for which there was neither precedent nor counterpart. He was a rare breed, a product of the American frontier juxtaposed with the new age of aerial technology in a remote and rural setting.

The pages that follow fall generally into two separate sections. One focuses on the Escobar Rebellion and Mexico's use of modern military aircraft in a combat environment; the other examines the United States Army Air Corps operations at the Johnson's Ranch airfield and subsequent aerial reconnaissance along the international boundary. All facets of the narrative are intertwined; the border experience deems it so. From this volatile political and military climate emerges the impetus for a number of decisions and operations that altered the course of Big Bend history.

In gathering the data for this book, I feel singularly honored to have known personally the two major civilian figures of this study, Elmo Johnson and W. D. Smithers. Each was remarkable in his own way, possessing a genuineness, a steadfastness, and a sincerity seldom met with in contemporary society. Both were products of a frontier environment, now long past.

I am indebted to Elmo Johnson for launching me on this project. He, unfortunately, never knew this; nor did he intend to do so. While interviewing him in June 1966 in preparation for the *Quicksilver* book, he mentioned casually that "you know that the Army Air Corps had a landing field on my ranch." Of course, I did not but was fascinated (my interest in flying began with Lindbergh's Paris flight). He continued: "Yeah, got to know a lot of fine people. 'Skinny' [Gen. Jonathan M.] Wainwright [hero of Corregidor and survivor of the Bataan Death March, also a Medal of Honor recipient] was at my home many times and 'Nate' [Gen. Nathan F.] Twining [former chairman of the Joint Chiefs of Staff] is one of my best friends." That did it! With one sentence this old West Texas rancher destroyed his credibility. As I began packing my briefcase to make a hasty exit, he tossed me an envelope. Embossed on the back was "Chairman, Joint Chiefs of Staff, Washington." The enclosed letter began "Dear Elmo" and was signed

"Nate." At that point, I realized Elmo was "for real." I unpacked my briefcase; Elmo talked Big Bend aviation while I wrote. At that very point *Wings over the Mexican Border* was conceived.

Two days later I met W. D. Smithers in Alpine, Texas, still gathering material for *Quicksilver*. When I mentioned my interest in "writing something" on the airfield at Johnson's ranch, Smithers was delighted. He also shared my long-standing interest in aviation, was instrumental in establishing the Big Bend airfield, and offered to aid me in researching the project. His contribution was invaluable; his kindness and friendship enduring.

But there were others, many, many others. Only through the contribution of those who remembered was this book possible. Because I was working with a topic that lacked an abundance of recorded data, oral sources were increasingly important in filling in the data gaps. To retired Air Force officers Col. John W. Egan, Maj. Gen. William L. Kennedy, Col. Don Mayhue, Brig. Gen. Robert S. Macrum, Maj. Gen. Hugh A. ("Lefty") Parker, Col. Elmer P. Rose, Col. Roy P. Ward, Lt. Gen. Samuel L. Myers (USA, Ret.), Rayma L. Andrews, and E. G. ("Buster") Ashley and his wife, Thelma, a very heartfelt thank-you for both your help and the lasting friendships that came from this experience.

An author preparing a historical manuscript is ever dependent on multiple data sources. In this instance, many people and many institutions made significant contributions; my bibliographic cup runneth over.

Libraries and archives: Patricia Finnell Scott, former assistant archivist, Archives of the Big Bend, Sul Ross State University, Alpine, Texas; Bonnie Grobar, Public Services Department, and Chris LaPlante, assistant archivist, Texas State Library, Austin, Texas; James N. Eastman, Jr., chief, Research Division, Albert F. Simpson Historical Research Center, Maxwell Air Force Base; Frances Rodgers, assistant archivist, Barker Texas History Center, University of Texas at Austin; Roy L. Flukinger, curator, Photography Collection, Harry Ransom Humanities Research Center, University of Texas at Austin; Bud Newman, director, Special Collections and Archives, University of Texas at El Paso; Michael P. Musick, Military Archives Division, and Ferris E. Stovel, Civil Archives Division, National Archives and Records Service; Col. Lester E. Hopper, chairman, National Historical Committee, Civil Air Patrol, New Orleans; and especially Jay Miller, whose extensive private aviation history library was always open for my perusal.

Correspondents sharing both information and photographs: Riley M. Barlow, John F. Kasper, Lois Neville Kelley, Tom Mahoney, David L.

Smith, Lt. Col. Fred R. Freyer, Col. George W. Hansen, Maj. Gen. Joseph H. (Jimmy) Hicks, Col. LeRoy Hudson, Maj. Kyle Johnson, Lt. Col. Howard E. Rinehart, Maj. Gen. P. H. Robey, Maj. Gen. Harry W. Johnson, and Allan Levine. Each contributed significantly to this project.

I am especially indebted to those who read the manuscript and gave critical and well-informed suggestions for revision. These include Professor Stanley R. Ross, Department of History, University of Texas at Austin; Professor Jim B. Pearson, Department of History, North Texas State University; James H. Doolittle (Lt. Gen., USAF, Ret.); and friend and neighbor Albert S. Hopkins, Jr., who provided both encouragement and critical evaluation (he also corrected my spelling!).

It is appropriate that I repeat here my message of appreciation to H. Gilbert Lusk, superintendent of Big Bend National Park, for authorizing Mike Fleming, resource management specialist, to escort me in a four-wheel-drive vehicle to the Johnson's Ranch site, located in a remote and now virtually inaccessible section of the park. This remains one of my more memorable adventures in historical research.

I reserve my deepest expression of appreciation for my wife, Janet. Because of her achievements in the world of business investment, I enjoy the freedom to write. Thanks to Janet, and Janet alone, this and a previous book were possible, as well as at least two others now in various stages of research. In addition to being my benefactress and wife of forty years, she shares the pleasures of my accomplishments, and I hers. What better prescription is there for a rich and rewarding life? Thank you, Janet.

Austin, Texas
October 1983

"flying. . . an exhilaration beyond description"

INTRODUCTION

In December 1903, two young men from Ohio assembled a strange-looking winged device on the North Carolina coast. Unknowingly, they were about to usher modern civilization into a new age of technology. In a series of wobbly flights, the longest lasting only fifty-nine seconds, Orville and Wilbur Wright had, in that inauspicious setting, freed humans from their earthbound shackles and converted their world from two dimensions to three (Bilstein, 1984). Strangely, this landmark event caused little excitement; only two of the nation's newspapers carried the story. And for more than two decades, including a world war in which combat aircraft played a significant role, aviation remained a marginal oddity, its commercial potential unrealized, its military usefulness still in debate.

In retrospect, the immediate postwar years emerge as a bulwark to aviation progress. Commercial application of aircraft development achieved during World War I remained obscured in the shift to a peacetime economy. But upon entering the century's third decade, the American nation altered its philosophical bearing.

The switch in national leadership that occurred in Washington, D.C., on March 3, 1921, symbolized the changing mood in America. The electorate, rejecting Woodrow Wilson and the responsibilities of world stewardship, appeared ready to savor the rewards of a world they helped make safe for democracy.

In their quest for something called normalcy, Americans embarked on the great escape. They craved excitement. In satiating their appetites they explored heretofore forbidden and untrodden paths, baring a frayed moral fiber to a world aghast at their excesses and exaggerations. While bootleggers and gangsters rendered Prohibition a failure and political corruption became a national scandal, the nation's youth launched its own rebellion, rejecting everything held sacred by an older generation still clinging to cherished "established values." From the antics of youth-on-a-spree came a new image and a new vocabulary: bobbed hair, bell-bottom pants, rumble seats, hip-pocket flasks, petting parties, and SEX! Novelist Warner Fabian termed the practitioners America's flaming youth; F. Scott Fitzgerald became their literary spokesman; unquestionably, they made the twenties roar. America was losing its innocence.

As the decade reached midpoint, however, a growing psychological paradox began to emerge. Flaming youth, seemingly bent on premature burnout, began to realize something had been lost in the process; their expectations remained unfulfilled. Social observers identified a yearning for the symbols of youth and innocence that many had allowed to slip from their grasp. America needed something to cling to, to identify with. It needed a hero image.

Politicians and military heroes were now considered passé. In their place arose a horde of personalities, their achievements compatible with the spirit of the age, their images adding a tenuous sense of stability to a generation adrift. Sports figures and movie stars emerged as the premier hero-celebrities of the 1920s. Babe Ruth, Red Grange, Jack Dempsey, Bobby Jones, Bill Tilden, and Knute Rockne all rivaled the president of the United States for name and image identification. So did the stars of the silent screen: Harold Lloyd, Charlie Chaplin, Fatty Arbuckle, Douglas Fairbanks, and Mary Pickford, "America's Sweetheart." In an age when all America succumbed to hero worship, they gave the 1920s an identity, a fleeting sense of image.

But the hero America had unknowingly been seeking arrived unexpectedly on a morning in May 1927. When the wheels of Charles A. Lindbergh's Ryan monoplane, "The Spirit of St. Louis," touched the surface of Le Bourget Field near Paris, the man, the airplane, and the event ignited a worldwide explosion of public adoration. Nothing like it had occurred before and nothing since, not even a man landing on the moon, has elicited that kind of response. In that one brief moment Lindbergh became a world celebrity, a hero to all. He was honored by kings, queens, princes, and presidents, and millions of people on two continents thronged to see the first man to span the Atlantic Ocean, alone in an airplane. Lindbergh was the man of the age, the fulfillment of the emotional and psychological buildup that had been fomenting since the end of World War I.

Although Lindbergh's hero-celebrity status lasted little more than a decade (his pro-German pronouncements issued in 1940 produced a strongly negative reaction), the impact of his achievement endured. Psychologically, his conquest of the Atlantic "became a metaphor for the mastery of the complexities of the twentieth century" (Bilstein, 1984: 161). Technically, this was the most important breakthrough in the annals of aviation. The world was suddenly smaller; a new age of commerce and travel awaited somewhere beyond the horizon.

Foreseeing the changes his flight portended, Lindbergh wrote: "The airplane has now advanced to the stage where the demands of commerce are sufficient to warrant the building of planes without regard to military usefulness. . . . Undoubtedly in a few years the United States will be covered with a network of passenger, mail and express lines" (Lindbergh, 1927: 136). His prophecy was soon realized; the trans-Atlantic flight ignited the aviation boom in America. "Suddenly in the summer of 1927," wrote aviation historian R. E. G. Davis, "the United States was caught in a wave of enthusiasm for aircraft and aviation which increased the momentum at an astonishing rate" (1972: 56). Almost overnight executives of infant airlines began ordering big trimotor airplanes (many began with single-engine aircraft) while most urban communities rushed to build airports to accommodate these new vehicles of travel. In 1928, the year following Lindbergh's historic flight, airline traffic quadrupled.

The unknown potential of aviation challenged men and women to test aircraft performance to its absolute limits. Distance became the first great test; this yielded America's first heroes of peacetime aviation. The Atlantic and Pacific oceans presented the supreme challenge to an aircraft's endurance; the greatest menace in case of failure offered the least chance of survival. In spite of the hazards, many came forward to challenge these formidable waters. Between 1919 and 1931, for example, sixty-three pilots and/or crews attempted North and South Atlantic crossings, either by airship (eighteen lighter-than-air units) or aircraft. Of the latter, thirty-five completed the crossing (*Aero Digest*, Dec. 1931: 33–37).

Prizes offered to generate interest in aviation account for much of this transoceanic air traffic. The twenty-five-thousand-dollar Raymond Orteig Prize for the first nonstop flight from New York to Paris and the thirty-five-thousand-dollar California-to-Hawaii Dole Prize did indeed stimulate interest in aviation. The challenge of transoceanic flight lured many entrants, yielded three winners, and produced some of aviation's darkest hours.

The bizarre achievements of these aerial stuntmen yielded positive benefits. During the late 1920s and the early 1930s, they demonstrated the potential of manned flight, which helped usher America and the entire industrial world into the modern air age. And even though the depressed economy of the early 1930s retarded expansion in the aviation industry, the public's interest in flying never slackened.

The cost of flying, which made it unobtainable for most civilians, further sharpened the spectator's appetite, as an expanding cadre of young men and women with wings continued to explore the frontiers of the air.

During the early 1930s, speed and endurance remained aviation's supreme challenges. Progress in engine and airframe design greatly increased performance capability, forecasting the airplane's ultimate potential. The National Air Races, held each autumn, became aviation's showcase of progress. The two-year performance record of Maj. James H. Doolittle dramatizes the rapid changes occurring in the aviation industry. In 1931 he won the transcontinental Bendix Trophy Race with an average speed of 209 miles per hour, while the following year, entered in the closed course Thompson Trophy Race, he established a new world's land plane record at 296 miles per hour, almost one hundred miles per hour faster than his speed the previous year. (At that time a Vickers-Supermarine S6B seaplane held the world's speed record at 407.5 miles per hour.)

Distance, however, re-emerged as aviation's primary challenge. Transoceanic flights had become passé; all that remained for the distance flyer was to race the clock while circling the earth. Between June 23 and July 4, 1931, Wiley Post, with navigator Harold Gatty, flew around the world in eight days, fifteen hours, fifty-one minutes. Their average speed in the Lockheed "Vega" monoplane "Winnie Mae," was 138 miles per hour. Two years later Post completed the same flight alone, reducing his previous time by twenty-one hours and establishing a new world's record.

The aviation activity most Americans witnessed firsthand occurred at the local municipal airport. This facility, most likely a flat pasture with minimum peripheral obstructions, became the focal point in local aviation activity as well as a source of civic pride. The establishment of a municipal airport constituted the pinnacle of urban achievement, an event that merited a spirited celebration. Invariably scheduled on weekends to insure maximum attendance, the dedication format included airplane rides (usually one dollar and up), air races, stunt flying, parachute jumps, speech making, and, if at all possible, a demonstration of formation flying and combat maneuvers by an Army Air Corps unit.

Lt. John W. Egan, who led a three-plane flight to the dedication of the Baton Rouge, Louisiana, municipal airport, remembered the official welcome accorded him and his flight crews. During a reception in the state capitol, the lieutenant governor asked Egan to perform some special aerial aerobatics and formation maneuvers during their exhibition, and, Egan explained, following the dedication ceremonies "we were to come back [to Baton Rouge] anytime we could and *always dive on the capitol building*. Make as much noise as we could. He wanted those engines roarin' so that the people of Louisiana would know what aircraft were, that they had an airport there, and that *Louisiana had entered the air age*" (interview, February 7, 1978).

America had indeed entered the air age, both literally and psychologically. Men in flight, regardless of the type of aircraft they flew and the maneuvers they performed, held spectators spellbound; airplane watching, horizon to horizon, became a national pastime. Caught up in the force of this emotional groundswell, many people, young men especially, wanted to learn to fly. And some who could afford the luxury invested in "flying lessons," readily available at varying degrees of safety and pedagogical excellence. Their objectives: adventure, fame, and fortune.[1]

Few found fame, still fewer achieved fortune, but adventure awaited the students of flight. Fixed-base operators located at most municipal airports offered dual instruction by the hour, half-hour, or whatever time interval the aspiring student pilot could afford. Those who could afford the tuition at the United States approved flying schools and air colleges received a higher-quality, better-organized, and usually safer instructional regimen. Tuition at these institutions ranged from $340 for a private pilot's course to $3,975 for a transport pilot's course at the T. C. Ryan school in San Diego, California. In 1931, the Dallas Aviation School offered a similar six-month, two-hundred-hour transport course for $2,500, which by 1934 had been reduced to $1,795. Although the aviation schools claimed success in securing employment for their graduates, relatively few could afford the cost of instruction.

For those wishing to embark on a career in aviation, the United States Army Air Corps provided the ultimate opportunity. Flight instruction in the late 1920s was conducted at two centers: March Field, California, and Brooks Field, Texas. However, with the establishment of Randolph Field in June 1930, all primary flight training activities were transferred to that facility. (Randolph Field was not fully staffed until October 1931.) Referred to as the "West Point of the Air," this base became the mecca for thousands of young men hoping to enter aviation through the military. Once admitted to the flight training program, the cadet studied a thoroughly comprehensive curriculum conducted by well-qualified instructors in the most modern aircraft and received a salary while learning to fly. Upon graduation, oppor-

---

1. During the 1932–34 biennium, an airline pilot's average annual income was $8,000, second only to a congressman's at $8,663, and more than double the $3,382 earned by a doctor (*This Fabulous Century*, 1969: 24).

tunities beckoned. The young officer could choose between a career in the Air Corps (providing appointments were available) or a lucrative position with the airlines, which gave preference to military-trained pilots. As might be expected, competition for admission was horrendous.

To be accepted for pilot training in the early 1930s, the Air Corps admitted only those individuals who possessed outstanding physical, mental, and emotional capacities. Comparatively few were accepted and still fewer survived the rigorous training regimen. Although a college degree was recommended, preferably with a major in engineering, an applicant with at least two years of acceptable college work could take a comprehensive examination for placement in a forthcoming class. Many applicants, qualifying academically, frequently faltered during the physical examination, and a 90 percent attrition rate was not uncommon.

Although those accepted by the Air Corps for flight training were outstanding physical and mental specimens, at least 50 percent usually failed to complete the course. Challenges to survival abounded. Col. Roy P. Ward, who graduated from the Advanced Flying School at Kelly Field in 1931, explained that "the instructors constantly evaluated your demeanor, your conduct, and how you could stand pressure on the ground. [And we were] subjected to constant pressure by our instructors" (interview, May 3, 1981). "Pass or fail," Gen. William L. Kennedy recalled, "we all had to suffer. They [the instructors] made us suffer." Remembering his cadet days, Ward added, "[They] were washing out a lot of good pilots. There just wasn't room for them" (interview, January 9, 1981).

Those who survived cadet training graduated as second lieutenants in the United States Army Air Corps. To these young officers, this marked an important milestone in achieving life's greatest ambition— to fly. General Kennedy summed up the feeling, "We all loved airplanes" (interview, July 30, 1981). But to the American public these flyers were society's elite, men with wings who flew the nation's airplanes bearing the red, white, and blue star insignia, symbolic of the republic they were committed to defend. Those who witnessed the massed fly-over on graduation day at Kelly Field stood in awe of the spectacle, their responses profound and widely universal. The pageantry of airplanes in neatly measured formations was inspiring indeed, evoking a feeling of latent patriotism in an era when the patriotic impulse deeply touched the souls of men and women. Many had lived through a world war and an economic catastrophe. They were searching for a more stable and fulfilling existence in the new era that aviation symbolized. Manned flight, still shrouded in mystery and highly romanticized, projected an all-encompassing panacea for a brighter future.

The public's approbation of these young men with wings also contained a large measure of hero worship. By the late 1920s, the images were well defined. Visions of the legendary fighter pilots of World War I, itinerant barnstormers, and adventurous airmail pilots, stimulated further by the Lindbergh Syndrome, remained a powerful impulse in American society. This did not go unnoticed by the young officers graduating from the Air Corps flying school. Colonel Ward, never modest in his appraisal of society's view of him as a second lieutenant, claimed with justified pride, "Hell, I was a hero and a celebrity too" (interview, April 9, 1980).

Celebrity status in 1931 carried a high price. And while danger came with the assignment, the opportunity to fly merited the risk. "It was a challenge," Colonel Ward explained optimistically, "but if you were properly trained the odds favored survival. . . . We gave hostage to the gods for the opportunity of learning to fly in the sense that we were put on notice that we might be required to make the supreme sacrifice." The margin of safety that Ward and his colleagues experienced was specifically circumscribed; they "were taught to respect those limits and not exceed them without knowing the penalty." To those who suffered the supreme penalty, the survivors justified their colleague's misfortune with the morbid rationale, "he died in his airplane." As one who survived, Ward concluded, "You developed a great respect for what you were doing" (ibid.).

To comprehend the impact of aviation on the American nation in the 1920s and 1930s, one must examine that experience in the broad context of world civilization. Within that vast time frame measured vaguely in hundreds, thousands, and millions of years, man had been free from earth's confines only thirty years when Roy P. Ward learned to fly. In less than three decades following the Wright brothers' achievement, Charles A. Lindbergh taught the world, suddenly and emphatically, that the airplane would alter the course of the human experience. And Lindbergh's triumph occurred just four years before Roy P. Ward graduated from flying school. Changes in the thoughts and habits of man were occurring at a staggering pace.

A half century after the historic New York to Paris flight, it is difficult to recapture the emotional sensation of a nation discovering wings. The recollections of those who appeared in the vanguard of that great adventure, however, explain that experience in terms of man's fervent desire to fly. Colonel Ward regarded flying as "a bright adventure to adventurous young men of that day. I must say in all honesty [we were motivated] by a sheer desire to just fly. . . . It was an exhilaration beyond description to get in an airplane, take off, climb to any height you wanted to and no longer be earthbound" (ibid.).

The pages that follow document the actions of these adventurous young men fulfilling that fervent desire to fly. The narrative focuses on two sectors: Mexico during the 1929 Escobar Rebellion and the United States–Mexican border during and following this encounter. Violations of United States territory during the closing phases of the rebellion produced a two-dimensional counter-operation. Air and ground forces assigned to critical locations along the Mexican border secured the lives and property of United States citizens while hastening the end of the revolt. (Significantly, this was the last combat-ready mission of official United States forces prior to that "day of infamy" in December 1941.)

Fear of a repeat of postrevolution border banditry led to the establishment of an Army Air Corps emergency landing field on Elmo Johnson's Big Bend ranch. During the 1930s and later, border protection became largely a matter of an aerial show of force. The conditions that developed around this riverside airfield attracted hundreds of daring young men in open cockpit biplanes to patrol the border, increase their aerial proficiency, and perform whatever aerial hijinks necessary to prove their mastery of aircraft. The blatant violations of Air Corps regulations that occurred in the process prompted one retired colonel to comment, "There was just nothing else quite like Johnson's Ranch."

The primary focus of this study is the military pilots and the airplanes they flew. This cadre of young aviators represented the new hero-celebrities of the air that had captured the imagination of the American people. Their image would continue to grow in scope and substance, ultimately to evolve as a new American folk hero. With helmet and goggles, white scarf and leather jacket, and parachute loosely harnessed to his upright frame, no other fighting man (save possibly the armored knights of the Middle Ages) had cast such an imposing silhouette. He symbolized a new era of international warfare; as military strategy changed, combat responsibilities shifted accordingly from the ground to the air. This was the forecast of things to come.

Although the Johnson's Ranch experience emerges as a strange mishmash of quasi-military, quasi-civilian activities, it translates into an introspective study of the growth pattern of military aviation in America. Against a background of international involvement and an airfield dominated by the unique personality of a Big Bend rancher and border trader, the following narrative examines in sharp focus the attitudes, the expectations, and the skills of some of America's unrecognized pioneers of military aviation. Within this microcosmic study is documented the efforts of young military aviators to demonstrate the viability of combat aviation; it had a future and, to them, the future was now. The ensuing course of world events allowed many of them to translate their beliefs into dramatic and victorious action.

"...a sentimental journey"

PROLOGUE

It began as a sentimental journey to revisit once-familiar sites and rec-
ollect those still-graphic images of the past. But this was not to be;
man and time had wrought their changes. The people, the voices, the
friendships—all the things that enrich and magnify the experiences of
youth—had long since disappeared. Only the land, the river, and the
memories remained.

And as most such journeys end, so did this, with a feeling of re-
morse for old friends once known and good times almost forgotten.
But to attempt to recapture and relive the past one first has to search
for those times and experiences that decades have thoughtlessly sealed
off from the present. And thus the past, sometimes seeming so near, is
relegated forever to wishes, to dreams, and to history.

And so it was when Wilfred Dudley Smithers, photographer, journalist, and historian, asked his old friend and colleague, J. O. Casparis,
bush pilot and aerial eagle hunter, to fly him from Alpine to the Big
Bend country that morning in 1962. Smithers had first witnessed the
raw, rugged beauty of that region in 1916 as a United States cavalry-
man. He chose to remain; that harsh, unyielding environment placed
an indelible stamp on his subsequent work. Readers were intrigued
with his accounts of primitive life along the Rio Grande, and his photographic renderings of the people, places, and events captured a disappearing way of life on one of America's last frontiers. This was his
country and he wanted to see it again. The past still seemed so near.

As Casparis climbed his 1941 Aeronca Chief monoplane to the cruising altitude, the two old friends watched familiar landmarks slip slowly
by—Cathedral Mountain, Santiago Peak, and Nine Mile Mesa. Following a southward course toward Mariscal Canyon, the point where
the Rio Grande carves its deepest penetration into Mexico, the old

eagle hunter guided his vintage aircraft around the eastern perimeter of the Chisos Mountains. As they passed over Dugout Wells, Smithers began looking for his first photographic checkpoint, the abandoned ruins at Glenn Springs, site of the tragic 1916 Mexican bandit raid.

The village site came into view, then slowly disappeared as they circled south of the Chisos range and began following the Rio Grande on a northwestward course toward Santa Elena Canyon. Approaching the point where Smoky Creek intersects the great river, Smithers pointed ahead and Casparis gently banked the Aeronca to the left, affording him an unobstructed view of the panorama below. Smithers raised his camera, snapped the shutter, and captured on film the fading site of his early Big Bend adventures—crumbling ruins of the Elmo Johnson ranch house and trading post, the dim depression that was once his dugout darkroom, and, stretching beyond, the abandoned airfield, one of the most remote and unique installations in the annals of the United States Army Air Corps.

This is where it had all begun over three decades before. Squadrons of the Air Corps' most advanced combat airplanes once lined the dim earthen runways, now overgrown with mesquite, greasewood, and wild tobacco plants. And to the west, on a high bluff overlooking the Rio Grande, stood the crumbling remains of the radio station. From this remote transmitter the United States Army intelligence once received daily dispatches on conditions along the Mexican border.

As Casparis circled the ruins, Smithers gazed down on the grim remains, reminders of a rich lifestyle he had once known with Elmo and Ada Johnson. Smithers had lived with these modern pioneer ranchers, traders, and humanitarians who played host to both the famous and infamous. Impoverished goat herders from across the Rio Grande, as well as people of wealth and eminence enjoyed the Johnson's hospitality. Bandits and lawmen, authors and scholars, scientists and government officials, cavalry soldiers and Air Corps pilots—all had been visitors at the Johnson ranch. Three great pillars of melting adobe, still visible on the west side of the dwelling site, marked the location of the broad patio that overlooked the Rio Grande. In this romantic setting famous generals disregarded rank and relaxed casually with lieutenants, sergeants, corporals, and privates. And all stayed for dinner to enjoy Ada Johnson's good home cooking. "Mrs. Johnson fed generals and privates at the same table," Smithers later recalled (interview, August 27, 1975).

Between the main house and the river below spread the dry barren floodplain where Johnson and his Mexican neighbors cultivated irrigated fields of fruits and vegetables. Smithers had known them all.

Johnson was his friend, as were the Mexicans who worked the fields. Smithers spoke their language, knew their names, and shared their folk secrets as did few outside their native culture.

After circling the area a few minutes, Smithers was ready to return to Alpine. The journey was nearly complete; scenes of three decades past had been revisited. And while the tangible evidence of an earlier lifestyle was fast disappearing, the memories lingered—exciting years when the Big Bend area was still a remote frontier, when political up-heaval in Mexico spread fear among the Rio Grande settlements, and when military aviation, still in its teenage years, joined the United States Cavalry to give the region its first permanent security. But memories, like the fading panorama Smithers watched from his aerial vantage point, were becoming distant blurs. The past Smithers re-membered bore scant resemblance to the present. There was no need to remain; the mission was complete. His two-sentence summation is revealing in its brevity: "We flew to Johnson's but did not land. I saw all I wanted from the air."

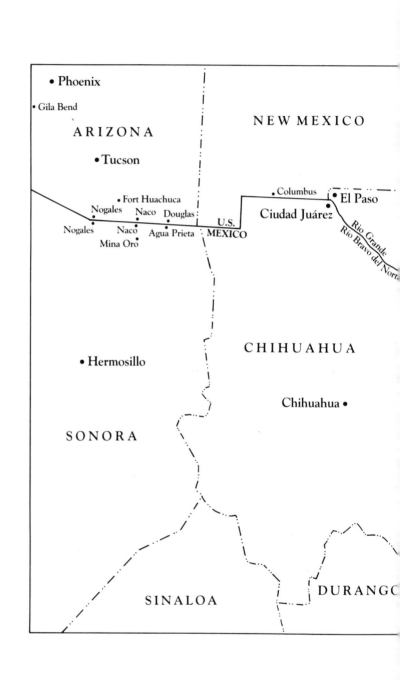

". . . presidential preference primary, old style"
—Duncan Aikman, "A General's War"

Scene 1.

In the Beginning, Another Mexican Rebellion

"...Manzo and Topete were fomenting trouble in Sonora"

CHAPTER I

The stability of time is frequently documented in nature, seldom in the works of man. Had Smithers looked across the Rio Grande adjacent to the Johnson ranch house—and possibly he did—he would have noticed one unchanged landmark, a great cottonwood tree that still cast its shadows along the river's edge. It was from beneath that tree on an April afternoon in 1929 that a band of Mexican horsemen, armed and trail weary, had gazed ominously across the river toward the ranch house and trading post on the bluff beyond. Elmo Johnson had watched the gathering with apprehension. He was well aware of what the assemblage portended.

Mexico's early twentieth-century history is punctuated with revolution. Díaz, Madero, and Pancho Villa symbolize that country's internal strife during the pre–World War I era. Four subsequent military encounters (1920, 1923–1924, 1927, and 1929) document Mexico's bloody trek to the ultimate institutionalization of the political process. Revolutions are seldom self-contained; people who lived on the Texas side of the Rio Grande involuntarily shared this experience. They lived with history and knew well the tragic lessons of the past.

Pancho Villa's attack on Ojinaga, the Glenn Springs raid, the Nevill Ranch raid, the Boquillas raid, Chico Cano's reign of terror—tragic symbols of Mexico's political instability—became watchwords along the Texas border. The remoteness of the region and its close proximity to Mexico tied it inseparably to the affairs of that nation, especially to its volatile political climate. And when Mexico's political leaders turned to armed rebellion to resolve their differences, those living in the Big Bend of Texas frequently felt the shock waves of battle. As Elmo Johnson looked across the Rio Grande on that April afternoon, he feared that history was about to repeat itself.

The assassination of Mexican president-elect Alvaro Obregón on July 17, 1928, had set in motion the chain of events that led to the afternoon tableau by the great river. The political vacuum created by Obregón's death reopened the way for political opportunism. Fighting flared throughout northern Mexico, producing instant replays of the all-too-familiar border clashes. In less than a year the backwash of rebellion again spilled over the international border, drawing the innocent and uninvolved into the conflict.

Elmo Johnson read well the signs of the times. He and his wife lived in isolation on the very perimeter of civilization, and if the expected attack should occur they alone would be forced to defend their lives and property. Yet time and distance shielded from the Johnsons one key element of this borderland drama: modern technology had invaded Mexico's revolutionary enterprise. Unknown to them, combat aircraft had been a factor in driving the renegade horsemen to their secluded hideout beside the Rio Grande. Also, the United States Army Air Corps possessed the attack capability to hasten the end of the conflict and ultimately bring a measure of security to the lonely couple that gazed apprehensively across the great river.

Frequent shifts in power are part and parcel of Mexico's political history, and with Obregón's untimely demise that country again faced a crisis of presidential succession. As the last surviving major figure of the Mexican revolution, Obregón, along with Adolfo de la Huerta and Plutarco Elías Calles, had led the armed revolt that toppled the Carranza government in 1920. During the ensuing decade Obregón had grown steadily in political stature, winning an overwhelming majority in 1928 and succeeding his former comrade-in-arms Calles as president of Mexico. In a nation accustomed to unconstitutional changes in national leadership, Obregón's demise again opened the way for political opportunism. President Calles, however, moved quickly to insure an orderly transfer of constitutional authority.

Refusing to serve beyond the expiration date (November 30, 1928) of his elected term, Calles urged the Mexican Congress to exchange political strong-man rule for a more stable governmental system based on laws and institutions. On September 25, 1928, the Mexican Congress unanimously selected Emilio Portes Gil, the former governor of Tamaulipas, as Mexico's provisional president. He was to take office on the following December 1, and a new president would be elected on November 20, 1929. As Calles' personal designate for the high post, Portes Gil was an ambivalent choice. In one respect he represented political continuity; in another, a break with the past. Although the interim president and Obregón were both Calles' political allies, Portes Gil was a civilian about to occupy Mexico's highest office, thereby breaking a cherished military tradition that the president be a general. "The fact that Portes Gil was not an army man," David S. Cullinane wrote, "was one of the causes of the rebellion, as the clique of army generals who had controlled Mexico for many years hated to see their power and prerogatives slipping away" (Cullinane, 1958: 98).[1]

Sensitive to the volatile climate Obregón's assassination created, Calles had solicited unity in the army in the succession matter and asked that no military man seek the presidency. Although the military heirs apparent had fully expected to select Mexico's next president from their own ranks, they agreed to abide by the congress' choice. Gen. José Gonzalo Escobar, whose loyalty Calles suspected, assured the outgoing president that "barrack uprisings have passed into history. The Army has been definitely purged of such shameless men" (Dulles, 1961: 389–390).

The president, for good reason, was not totally convinced. He recognized in Escobar's vacant rhetoric a resort to temporary expediency. Both Calles and Escobar were veterans of many political and military struggles and the outgoing president understood fully the power psychology that dominated Mexico's military establishment. From Calles' perspective, Escobar and other ranking army officers constituted the unstable element in the immediate scheme of things. But Gen. José Gonzalo Escobar was something definitely unique in the military hierarchy.

The dashing young moustachioed *galán* first achieved national prominence in 1919. As commander of the federal garrison at Ciudad Juárez, he led a successful cavalry charge against Pancho Villa. In driving the rebels from the city, Escobar sustained a bullet wound in his lungs. Remaining in his saddle, he rode to the international bridge, saluted the United States border guards, and informed them of his wounds. According to one report, he was taken to the Hotel Dieu where American army officers from Fort Bliss honored him for his courageous victory. For his daring exploit, President Carranza promoted Escobar from colonel to general.

Greater achievements still lay ahead. He emerged as the hero of the "Shelf of Death" battle that ended the de la Huerta revolt, and two years later helped suppress the abortive Gómez-Serrano revolt. For ridding Mexico of these "shameless men," Escobar returned to the nation's capital for a hero's welcome, riding triumphantly through Mexico City, leading his troops on a symbolic white horse. For helping quell the insurrection, Calles accorded Escobar the rank of divisional

---

1. Portes Gil possessed one additional disadvantage. Like Obregón, he was an *agrarista*. This created dissension among the ranks of labor.

general, placing him in command of something like ten thousand troops, a lot of power for a Mexican general (*Diccionario*, 1976: 716).

All of Escobar's conquests were not conducted on the battlefield. He frequently succumbed to temptations far removed from military politics. The handsome young general, now a national hero, found eagerly awaiting beautiful young women wherever his responsibilities carried him. According to one source, some loved the general with careless abandon, a candid explanation of why at one time there were three sets of Escobar children living within a few blocks of each other in El Paso.

The young general conducted his amorous conquests through arrangements, understandings, marriages, and divorces, but unfortunately the latter two never seemed to balance numerically. For example, the last Señora Escobar, Concepción Goeldner, a former Texas beauty-contest winner, claimed that the general had neglected to divorce the previous Señora Escobar, one Elsia Valles. In explaining Escobar's casual regard for marriage without benefit of divorce, one journalist wrote:

> There had been many unfortunate divorces in his career and, as hundreds in El Paso knew, he contributed regularly and sometimes lavishly to the support of all three families. If General Escobar had neglected to secure a divorce from Elsia Valles before marrying Concepción Goeldner, his friends say that the matter soon was remedied in Mexican courts where generals, at least, can secure decrees with amazing promptness on the most slender grounds. ("How General Escobar Fought," 1931: 2)

By the late 1920s, Escobar's popularity and military rank gave him the personal stature to make his bid for power should the prospects appear favorable. Of this, Calles was acutely aware. Historian René A. Valenzuela explained: "By the time Calles became president Escobar had climbed high on the list of potentially dangerous generals whom Calles pampered with financial opportunities of various sorts" (1975: 20–21). Calles knew from long personal experience that power and wealth beget more power and wealth, and within the context of Mexico's military tradition the temptation to perpetuate the cycle through force was ever present. With Obregón's death interrupting the mandated process of presidential succession, a new temptation awaited those equal to the challenge.

Escobar and his military associates were careful to conceal their displeasure with Portes Gil's appointment. Displaying overt gestures of support, Escobar reassured the new provisional president that the army remained the true defender of the Republic. Yet behind the facade of friendship and goodwill an army rebellion was brewing. United States ambassador Dwight W. Morrow's close personal relationship with Calles enabled him to keep his government apprised of developing trouble spots along the United States border and of the personalities who would probably emerge at the forefront of the movement. On February 19, 1929, Morrow reported to the State Department in Washington that Gen. Francisco R. Monzo, Governor (and Gen.) Fausto Topete, and Col. Ricardo Topete, "all warm partisans of General Obregón, who felt that they had not received proper consideration by the Provisional Government, were fomenting trouble in Sonora." The ambassador added that it was considered likely that General Escobar would join the revolution.[2]

In early February 1929, Portes Gil began preparing for the inevitable. He launched a two-phase plan of operation that solicited virtually unlimited United States support. Portes Gil requested Washington to seal the border to potential rebel needs while expediting the flow of war matériel to restock the federal arsenal. This all had a familiar ring. In March 1912, United States president William Howard Taft had proclaimed a general embargo on the shipment of arms to Mexico, and six months later, under executive order, allowed two airplanes to be exported to the federal forces to use against the rebels in Chihuahua. And again in 1924 the United States "accommodated President Obregón by imposing a new embargo and applying it only to the rebels" (Newton, 1970: 107, 118).

As expected, the Mexican government received carte blanche approval of both requests. In response to Acting Minister of Foreign Affairs Genaro Estrada's request, Ambassador Morrow alerted his government on the morning of February 11 to the possible shipment of arms and munitions through United States ports of entry consigned to destinations in the state of Sonora. The following afternoon Secretary of State Frank B. Kellogg instructed officials of the Treasury and Justice departments to "exercise special vigilance with the objective of preventing possible attempts to smuggle arms or munitions into Mexico" (Foreign Relations, 1943: 3: 336).

In increasing his arms inventory, the only problem facing Portes Gil was the immediate acquisition of matériel from United States manufacturers, but Morrow and the State Department again quickly provided a solution. (Military funding posed no problem for the interim president: he had inherited an abundant national treasury from his

---

2. *Papers Relating to the Foreign Relations of the United States*, 1943: 3: 419. Hereafter cited in text as *Foreign Relations*.

predecessor.) Through his acting minister of foreign affairs, Portes Gil advised Morrow on February 14 that his government planned to purchase combat aircraft as well as other war matériel from private firms in the United States and requested that the United States government facilitate these purchases. Secretary of State Kellogg took charge of the matter, advising Morrow two days later that "the Department would expedite the issuance of export licenses covering military supplies for the Government of Mexico." The secretary also conferred with Edward P. Warner, assistant secretary of the navy, who authorized the Mexican government to purchase military aircraft directly from a supplier already under contract to the United States government.[3] In the February 16 telegram, Kellogg informed Morrow that

> in view of the special circumstances in Mexico, [the Navy Department] would permit Vought [Chance Vought Corporation] to give preference to the Mexican order for either six or nine planes as Mexico might desire. . . . Machine guns and bombs, about which Colonel MacNab [American Military attaché in Mexico] telephoned, must be obtained through the War Department, which cannot make definite commitment until make of machine guns and exact size of bombs are known. (*Foreign Relations*, 1943: 3 : 337–338)

The Mexicans accorded the aircraft purchase high priority. Morrow received Kellogg's telegram approving the transaction in the early afternoon, and later that same day Gen. Juan F. Azcárate of the Mexican air service left Mexico City for Washington to complete arrangements for the immediate purchase and delivery of nine aircraft. The ambassador explained further that "the type of machine guns and bombs desired by the Government of Mexico are those for which the Corsair airplanes are equipped" (ibid., 3 : 338).

The United States Navy Department (contractors with the Vought firm), spurred on by the State Department, expedited the entire transaction—manufacturing, pilot training, and delivery. According to Chance Vought Corporation historian Gerard P. Moran, the first three O2U-2M "Corsairs" were delivered "less than two weeks from the date of order." And while the aircraft were still in production at the East Hartford, Connecticut, plant, navy instructors at Mitchell Field, Long

---

3. There was precedent for this action. During the de la Huerta rebellion of 1923–1924, the United States government honored the Mexican government's request to purchase United States military planes (Newton, 1978: 83).

Island, conducted an intensive training program for the Mexican aviators assigned to fly the "Corsairs" (Moran, 1978: 54). (Several different orders covering various delivery dates account for the total number of "Corsairs" purchased by the Mexican government.)

Following the subsequent outbreak of hostilities, the Mexican government increased the original aircraft order. According to Moran,

> Brig. Gen. Juan F. Azcárate, Chief of the Mexican Air Service, took delivery of the twelve O2U-2Ms at Mitchell Field in March. Then, in an unprecedented move, he personally led the planes south "for immediate employment against revolutionaries." Landing field assistance from New York to Mexico had been organized by the State Department and Navy. The twelve aircraft were the first armed "warplanes" of another nation to fly over United States soil. (Ibid., p. 54)

This transaction did not go unnoticed by the Escobar rebels. When a later flight of three "Corsairs" reached Texas, it was reported that the rebels had offered an agent ten thousand dollars to destroy the planes while at Love Field, Dallas. As a precaution against sabotage, a United States air reserve unit guarded the aircraft during the overnight stop in Dallas (*San Antonio Express*, April 16, 1929).

The night passed without incident, and shortly before one o'clock the following day the planes arrived at Kelly Field. During the brief stopover, a delegation headed by Gen. Frank Lahm, senior Air Corps officer at Kelly Field, and the Mexican consul at San Antonio greeted the ferry crews. While on the ground at the San Antonio area base, Air Corps mechanics discovered a leaking gasoline feed line on one of the aircraft: they repaired it while the flight crews lunched. Forty-five minutes later they were again airborne "en route to Mexico where they [the "Corsairs"] will be used by the federal government in its fight against the rebels" (ibid.).

In selecting the Vought "Corsair" to strengthen his military arsenal, Portes Gil chose wisely. These two-place all-purpose biplanes had already compiled an outstanding operational record with both the United States Navy and Marine Corps. Because of its combat versatility, China, Cuba, and several Latin American countries had also selected this aircraft to supplement their military air fleets. A 450-horsepower Pratt and Whitney "Wasp" engine (horsepower varied according to model) gave the aircraft a maximum speed of 151 miles per hour. As the principal offensive aircraft in these foreign air forces, the "Corsair" "fulfilled functions of pursuit, observation, bombing, and attack aircraft. . . . In its various configurations, the Corsair was well suited for these situations" (Moran, 1978: 54; Gray, 1929: 319c–320c).[4]

The battle lines were being carefully drawn. Again it would be Mexican versus Mexican, with the United States supporting the federal cause while pursuing a tenuous course of marginal neutrality. As in previous encounters, the name of the revolutionary game was supplies, United States supplies, which, for whomever's use, must be drawn through the ports of entry. Many of these key land access points, stretching from Brownsville, Texas, to Nogales, Arizona, would become the settings for violent military action, both ground and air. This would again place the thrust of revolution at America's doorstep, evoking dark memories of the previous decade.

Ambassador Morrow was indeed prophetic in designating the state of Sonora as the focal point of rebel unrest. That northwestern border state was traditionally one of Mexico's centers of political dissent. Native sons Calles, Obregón, and de la Huerta, "El Triángulo Sonorense" (the Sonoran Triangle), launched their campaign from there in 1920 to overthrow the Carranza government. Now, nine years later, another group of Sonoran rebels was organizing a similar campaign, proclaiming its guiding principles in the "Plan of Hermosillo." (The 1920 rebellion operated under the "Plan of Agua Prieta," having been organized in that northern Sonoran border village.) This document called for the people of Mexico to revolt and accused ex-president Calles of using bayonets, crime, bribes, and grants to thrust into the presidency Emilio Portes Gil, "one of the members of his [Calles'] troop of comedians" (quoted in Dulles, 1961: 438). The plan designated the rebels' crusade as the Revolución Renovador (Renovating Revolution) and recognized Citizen General of Division Don José Gonzalo Escobar as the supreme chief of the movement.

On March 3, approximately two weeks following the original aircraft order, the rebels launched their long-awaited attack on the federal government. Escobar maintained his charade of loyalty until the very end. Earlier that day he sent a telegram to Portes Gil placing himself at the president's disposal and then immediately led his troops in an attack on Monterrey. Two days later the capital of Nuevo León was in rebel control. Escobar's conduct of the Monterrey campaign established a pattern he would follow throughout the rebellion. "Escobar stayed in Monterrey long enough to take money from the banks there

---

4. Predecessors of the O2U-2M "Corsairs" had previously seen service in Mexico. In 1924, the USS *Richmond*, cruising off Veracruz, Mexico, launched a VO-1 Vought scout plane on a reconnaissance mission in support of Mexican president Alvaro Obregón's struggle with rebel leader Adolfo de la Huerta (Moran, 1978: 30–31).

[$345,000] (as he also did from Torreón banks) and to ravage the home of [federal general Juan Andreu] Almazán," Dulles wrote. "Then he retired in the direction of Saltillo, tearing up intervening railroad track" (ibid., p. 443).

The president moved quickly to counter the rebel uprising. In the absence of the ailing Joaquín Amaro, Portes Gil named Calles the interim minister of war and navy and placed him in charge of the military campaign against the rebels. With Monterrey already in rebel control, Calles' first objective was to intercept Escobar's army at Saltillo, capital of Coahuila. At this point, however, considering rebel defections, Calles was not certain of the number and location of the loyal troops in his command or the extent of the rebel following.

At the outbreak of the rebellion the leaders claimed they controlled ten states: Sonora, Sinaloa, Durango, Coahuila, Nayarit, Zacatecas, Jalisco, Veracruz, Oaxaca, and Chihuahua. In each of these states the switch from federal to rebel ranks followed a similar pattern. The military commander or governor, usually a general, defected from the regular army, in most cases taking with him his entire command. By the time Escobar attacked Monterrey, the rebels could claim a fighting force of almost thirty-thousand well-equipped men who had defected from the army.

Calles, on the other hand, found that he could count on about 72 percent of the original ground forces, the air force, and a division of about five thousand agrarians organized in San Luis Potosí on behalf of the federal cause. The challenge facing Calles, however, was formidable; his areas of command were vast and widely separated. Fighting would occur throughout much of the northern half of Mexico with most of the action taking place in two principal theaters of war: one consisting of the states of Nuevo León, Coahuila, Durango, and Chihuahua; the other Sinaloa and Sonora. Significantly, four of these six states bordered on the United States, a fact that would have vast international repercussions.

The rebels, like Portes Gil, recognized that victory was contingent upon United States support, but they sought it in a far different, if not a somewhat ill-conceived, manner. They believed the confusion and indecision surrounding the changing of administrations in Washington would be their greatest ally. By launching their attack on the eve of Herbert Hoover's inauguration, the rebels hoped, through a series of quick victories, to gain control of the Mexican government and thereby receive perfunctory recognition by the new administration whose foreign policies remained to be defined. This did not occur. "In the first place," Cullinane wrote, "Portes Gil . . . acted with great speed to suppress the revolution; and secondly, President Hoover

acted with unexpected promptness and firmness in support of the constitutional government of Mexico" (Cullinane, 1958: 97). Syndicated
columnist Drew Pearson, citing how Hoover's quick decision altered
the course of the revolution, wrote:

> All during inaugural day [March 4, 1929] when the decision of
> Herbert Hoover remained unknown, they (the rebels) won over
> garrison after garrison. But at noon of the day he entered the exec-
> utive office, President Hoover announced that the Coolidge policy
> of supporting the Mexican Federal Government would continue.
> From that moment the troops began to desert back to the Federal
> Government and the end of the revolution was in sight.

Pearson states further that Ambassador Morrow, who had cooperated
with the Calles–Portes Gil government, "is convinced that this government deserves to remain in office and he was instrumental in bringing Hoover to this point of view" (*San Antonio Express*, March 10,
1929).

Although Pearson's analysis proved essentially accurate, he was prematurely optimistic about an early end of hostilities. The war would
continue for weeks, more than two thousand people would die, the
nation would suffer severe economic loss (about forty million pesos),
and vast destruction of public property (railroads especially) would occur in the wake of battle.

The journalist's reference to Ambassador Morrow's role in the matter
reflects an even deeper insight into the ultimate outcome of the rebellion. In soliciting Hoover's support of the Portes Gil government,
the United States ambassador emerges as one of the revolution's pivotal figures by redirecting his country's support of the federal cause.
Through his enlightened understanding of the fundamental issues sep-
arating the warring factions and his deep sense of mission, he was able
to articulate accurately and without bias Mexico's urgent needs in a
time of crisis. Morrow's embassy communications to Washington,
therefore, reaped huge dividends for the government forces. And as
the high tide of revolution threatened to spill over onto United States
soil, this nation was drawn closer and closer to the conflict, further
strengthening Portes Gil's grip on the national emergency. In addition
to combat aircraft, the United States provided bombs, machine guns,
rifles, cartridges, and accessories valued at $480,908.94 (*Foreign Rela-
tions*, 1943:3:428–429).

It was, however, the introduction of the United States–built Vought
"Corsair" airplanes that helped give the Escobar Rebellion its unique
identity and became one of Portes Gil's major military assets. The

"Corsairs" gave the federal forces a flexible destructiveness unmatched by the enemy. They served as fighters and light bombers, carried out low-level strafing attacks, and became an important factor in altering the course of the war. The United States would also dispatch both its ground forces and combat aircraft to threatened border areas to help contain the civil strife within the boundaries of Mexico. The conflict was, therefore, destined to evolve as a distinctly different type of revolution with the airplane playing an unprecedented role in military tactics along both sides of the international border.[5]

---

5. Mexican revolutions provided an early proving ground for combat aviation. In February 1911, the Mexican government engaged René Simon, a member of an aerial circus touring the southwestern United States, to reconnoiter rebel positions near Juárez, Mexico. Subsequently a number of foreign mercenaries flew military-related missions for both Mexican federal and rebel armies. John Hector Worden was the first mercenary to fly in the Mexican Revolution. In 1912 he flew scouting missions for the Mexican government, occasionally dropping bombs on rebel forces threatening to dynamite railroad lines. In 1913 Dider Masson, a French-born American, piloted a rebel aircraft on several unsuccessful bombing runs on a federal gunboat in Guaymas harbor. That same year two American mercenaries, Dean Ivan Lamb and Phil Rader, representing opposing factions, engaged in what is thought to be the world's first aerial dogfight near Naco, Sonora. Their weapons: hand-held revolvers (Newton, 1970: 108–109; Obregón, 1917: 166–167; Dulles, 1961: 7; Seagrave, 1981: 6–30).

"... they stormed this office
seeking refuge in this country"

CHAPTER 2

In early March 1929, the American nation suddenly realized there was a war in progress at its very doorstep. And when America's hero-idol, Col. Charles A. Lindbergh, was reported near the war zone with his fiancée Ann Morrow, daughter of the United States ambassador, the nation's newspapers gave added coverage to the Escobar Rebellion. With combat forces taking up positions along the Texas and Arizona borders, it appeared the conflict would ultimately spill over the nation's southwestern perimeter. Almost overnight little-known desert villages such as Nogales, Naco, and Agua Prieta became familiar household words. Control of these ports of entry by the contending Mexican factions would emerge as the supreme prize of war.

Following the first day of the conflict, the rebel forces held the border towns of Agua Prieta and Nogales, Sonora. At Agua Prieta the garrison revolted against the federal government of Mexico and placed the collector of customs and several other customs officials in jail. According to United States Customs Inspector D. C. Kinne, "Not a shot was fired and there was no movement of refugees to this side of the border."[1] Nogales became another easy rebel victory. Late in the afternoon of March 3, about five hundred rebel soldiers arrived at this sister city of Nogales, Arizona, and

> placed the Collector of Customs, the Immigrant Inspector in Charge, the Post Master and the other chiefs of the Federal Service . . . together with all of their subordinates, under arrest and the soldiers took over the duties of such officials. . . . Many of the

---

1. March 4, 1929, Record Group 85. Hereafter referred to as RG.

Customs and Immigration officials swore allegiance to the Revolutionary Government and returned to their posts of duty at 8 A.M.
(R. M. Cousar, March 4, 1929, RG 85)

The capture of Naco, Sonora, followed a similar pattern. At three
o'clock on the morning of March 4, the revolting Mexican troops took
possession of the town. "Their main objectives appeared to be the
Customs and Immigration offices at the International line," United
States Customs Inspector Walter F. Miller recalled, "and as they advanced in skirmish order on these buildings, the greater part of the
Mexican federal employees retreated to the United States side of
the line, bringing with them their arms and ammunition." Miller reported that his office retained thirteen Mexican federal employees and
officials and their families pending instructions from Mexico City
(March 5, 1929, RG 85). Victims of war always include the civilian
population; the threat of the Naco situation escalating into open warfare sent crowds of civilians scurrying for safety. The customs inspector
described a colorful yet pathetic scene that would be repeated many
times all along the border.

The threat of an immediate rebel attack alarmed the civilian population of Naco, Mexico. . . . they stormed this office seeking refuge in this country. They resembled refugees from some storm or
flood stricken area, with their blankets, household goods, cooking
utensils and in most cases, parrots, chickens and the ever present
dog. (May 17, 1929, RG 85)

The quick successes at the Arizona ports of entry sparked renewed
confidence in the rebel movement. Even as United States troops began taking up positions along the international boundary, garrison
after garrison adjacent to the Texas border revolted, proclaiming allegiance to the rebel cause. On the night of March 5, troops from Villa
Acuña, Piedras Negras, and Ojinaga began a mass exodus as post
commanders led the defectors to General Escobar's aid in the central
Chihuahuan sector. According to Chief Patrol Inspector Chester G.
Courtney at Marfa, "General De La Vega . . . left with 500 soldiers in
the direction of Chihuahua City, leaving a Colonel in command at
Ojinaga." Foreseeing the deteriorating conditions this change in command would create along the Rio Grande, he wrote:

It is believed that when the different detachments of Revolutionary soldiers are placed up and down the river, with all the author-

ity that usually goes to bands of this class, they will soon start stealing from the Mexican population on the Mexican side, resulting in an influx of Mexicans to this side. . . . Mr. Blackwell also informed the writer there were fifty American refugees en route from Chihuahua to Presidio. It is expected by him more Americans will come out this way. (March 8, 1929, RG 85)

The opening scenes of the Escobar Rebellion, as well as those that would follow, sparked many unpleasant memories. To local citizens it appeared as a delayed rerun of a tragidrama that had played the territory a few seasons back. It had the same plot, same cast, and the same setting; only the names of the stars were different. Even the town names rang familiar: Naco, Nogales, Juárez, and Ojinaga. It had all happened before and now it was happening again.

Death and violence were inherent phases of border life and no one, neither Mexican nor Anglo, citizens of Sonora or Chihuahua or Texas or New Mexico or Arizona, was immune to the specter of revolution. And while the physical damage and destruction could be repaired with the cessation of hostilities, the psychological scars would persist along the borderlands long after the last shots were fired. These would remain visible in the decades to come. General Escobar had deemed it so.

Even before the fighting began, the rumblings of revolution were felt along the Texas side of the border. The civilian population followed Mexico's smoldering political crisis through the newspapers, while military units stationed throughout the region received periodic intelligence reports alerting them to the pending emergency.[2] Camp Marfa (later renamed Fort D. A. Russell), a cavalry post located adjacent to Marfa, Texas, had a long history of monitoring border disturbances in the Big Bend area. Established in 1911, various cavalry units were assigned there to protect that vast region stretching from the mouth of Boquillas Canyon to Hot Wells, located on the Rio Grande due south of Van Horn, Texas. Marauding bands of Mexican renegades made periodic raids on Big Bend ranches and villages, most occurring in the wake of Mexico's various political crises. For more than a decade that country's internal instability gave this remote cavalry post a critical role in maintaining border security. Since the mid 1920s, however, the region had been relatively free of border incur-

---

2. "Due to the unforeseen necessities incident to the recent revolution in Mexico, the maneuvers of the First Cavalry Division, planned as described in the April *Journal*, were postponed" (*Cavalry Journal* 38 [July 1929]: 436).

sions. This was reflected in the day-to-day activities at Camp Marfa. Unit enlistments had declined, and post assignments consisted mainly of routine training drills, post maintenance duties, weekend polo matches, and an occasional patrol mission along the Rio Grande.

When 2d. Lt. Samuel L. Myers graduated from West Point in 1928, his first duty assignment was the First Cavalry Regiment at Camp Marfa. At that time six troops (four line troops, a headquarters troop, a machine gun troop) and a band constituted the regiment; normally there would be fourteen troops. The troops were also at greatly reduced strength, each reporting about one hundred men and about 115 horses and mules. With these peacetime staff reductions, regiment strength in 1928 totaled approximately eight hundred men and one thousand animals.

No organized program of border surveillance was in effect when Myers arrived at Camp Marfa. "It was conducted in a rather hit or miss fashion," he remembered. Each of the troop commanders was "expected to send some kind of a patrol down into his sector [there were four in the regimental command] at least once a year. And that's about what they did." Some would take the entire troop on a two-week practice march, while others would "send a platoon down into the area to make a swing through it and back. . . . And I don't believe any serious, coordinated patrol was conducted during that period 1923 to 1928" (interview, 1980). This was destined to change.

Sometime prior to the outbreak of hostilities (Myers believed it was January 1929), each cavalry troop received orders to make a patrol, either in platoon strength or troop strength, in its sector each month. After the fall of Monterrey on March 5, the regiment command stepped up the border surveillance program by establishing cavalry outposts throughout the Big Bend area. These were located at Porvenir, Ruidoso, Presidio, Lajitas, Boquillas, and Castolon. From these outposts each troop patrolled thirty miles of the Rio Grande each day, marching fifteen miles upriver and fifteen miles downriver until they met the patrol from the adjacent sector.

The first unit from Camp Marfa to be ordered to the border on emergency assignment was "B" Troop. When the news reached the post that the Ojinaga garrison had defected to the rebels, Capt. Frank Bertholet, commander of "B" Troop, received orders to proceed immediately to Presidio and take up positions adjacent to the international bridge. Intelligence reports indicated that the Mexican federal army would probably attempt to retake the garrison in the near future. In case fighting broke out at Ojinaga, the captain's responsibility was to see that all combating parties remained south of the border.

On the morning of March 3, 1929, Captain Bertholet embarked on the sixty-six mile, two-day march with his 112-man force. A member of this unit was Pvt. Samuel G. Smith, who had volunteered for cavalry service only three weeks before "B" Troop began the march to Presidio. The son of a West Virginia coal miner who had never ridden a horse with a saddle, Smith chose the cavalry because he "wanted to get out into the western part of the country. And it seemed like the cavalry was always the most adventuresome part of the Army." As he began the march to Presidio, Smith feared he was about to receive far more adventure than he had bargained for. "When we left the post," he recalled, "we thought there was a possibility we were going into combat" (interviews, January 30, 1980; January 22, 1981).

As Captain Bertholet led his troops south toward Presidio, history was indeed repeating itself. Fifteen years before another cavalry force had embarked on a similar mission. Political instability had thrown Mexico into another military rebellion and, as in 1929, the borderlands region became the focal point of violent military action. On December 16, 1913, "A" and "C" Troops, Fourteenth Cavalry Regiment, and "E" Troop, Fifteenth Cavalry Regiment, established a temporary outpost in Presidio. With the repeated border clashes between the Constitutionalist and federal forces, their mission was the same as Bertholet's: contain the fighting south of the Rio Grande.

Early in 1914, the Constitutionalist forces broke through the federal defenses at Ojinaga, sending the defeated army, along with their equipment, supplies, horses, families, and followers, scurrying for safety. Texas offered their only route of escape. Arriving in Presidio, the Mexican general officers surrendered to Maj. Michael M. McNamee and the soldiers (approximately five thousand) were then interned as prisoners of the United States. On the morning of January 16, 1914, the cavalry began the return march to Marfa, escorting the walking army through Texas on a sixty-six-mile "trail of tears." Heat, hunger, and illness—many were wounded—added to the misery of the pathetic trek. Major McNamee, commander of the Fifth Cavalry Regiment, reported:

> there was no water along the road for the first fifteen miles and at this point an unhealthy water hole was found. When about twelve miles from Presidio one woman gave birth to a child. The child was dead and had to be buried near the road side. These conditions caused great hardships and suffering and resulted in a lengthening of the column for many miles. . . . All but about 200 reached Shafter by 10:00 P.M. Those unable to continue the march were

permitted to camp along the roadside. . . . About sixty animals either died or had to be abandoned along the road.[3]

Thus the tragedy of rebellion reached Texas soil. The purpose of Captain Bertholet's mission was to insure that this did not occur again.

When Bertholet's troops arrived at Presidio they were billeted in the old army barracks north of the village and about a dozen men were assigned to the United States Customs building near the international bridge. The automatic rifle platoon mounted their weapons on the roof of the customs building, sighting their guns through the openings in a decorative parapet that encircled the structure's roof. This position gave the soldiers protection from possible enemy fire while affording an unobstructed view of the target area. "From this spot," Smith explained, "we could cover anything from the Customs house to the international bridge" (interview, January 22, 1981).

Captain Bertholet found the military situation in Ojinaga stable, with normal traffic flowing across the international bridge at Presidio. A few days later—Smith recalled it was on a Sunday morning—fighting broke out on the outskirts of Ojinaga and lasted for about two hours. While members of the troop could not see the action, the sound of gunfire was clearly audible from the customs house and smoke could be seen rising above the scene of battle. "After about two hours all was quiet again," Smith explained. "We never did learn who won the battle" (ibid.).

In the wake of the fighting a number of Mexican refugees crossed to the Texas side of the border and were interned by the troops. The sight of the incarcerated refugees made a lasting impression on young Henry Aplington II, who visited his father, company commander Maj. Henry Aplington, at Presidio. "I can remember going down there from Marfa to visit the camp," he wrote. "The most impressive thing to a youngster was a wired enclosure where Dad had a bunch of refugees interned—big hats, serapes and all. Quite a sight" (correspondence, January 9, 1980).

---

3. McNamee to commanding general, Southern Department, February 10, 1914, RG 94. See also Lona Teresa O'Neal Whittington, "The Road of Sorrow, Mexican Refugees Who Fled Pancho Villa through Presidio, Texas, 1913–1914," master's thesis, Sul Ross State College, 1976; "Shot and Shell, The Fall of Ojinaga in 1914," *Voice of the Mexican Border* 1, no. 1 (September 1933), 13–21; and Michael Meyer and William L. Sherman, *The Course of Mexican History*, 2nd edition, New York: Oxford University Press, 1983.

Following the end of the fighting in Ojinaga, Captain Bertholet dispatched several patrols up and down the Rio Grande in search of renegade Mexican soldiers who might have escaped across the border. None were sighted, and, contrary to Smith's earlier expectations, no shots were fired during the Presidio assignment. "B" Troop remained in Presidio until April 6, when it was relieved by Lt. Lawrence G. Smith's "A" Troop.[4] For Aplington the sight of the cavalry troop embarking on the return march was another unforgettable experience. He would write later: "We will never see again the picture made by horse cavalry going into the field, the officers and troopers with their field saddles, guidons flapping, machine gun horses clanking, and . . . the whole column followed by the escort wagons" (correspondence, January 9, 1980). During the coming weeks, however, the sight of cavalry troops taking the field in combat readiness would be repeated frequently throughout the Big Bend country as the United States military responded to the Escobar emergency.

While "B" Troop monitored conditions in the Ojinaga-Presidio sector, Lieutenant Myers received orders to take his platoon—thirty men and thirty-two horses—to Porvenir, a rural settlement on the Rio Grande about thirty miles south of Valentine, Texas.[5] At this location Myers was to set up a base camp, patrol his thirty-mile sector each day, and report any troop movements observed along the Rio Grande. Although the regiment provided for the platoon's ordnance requirements (each man carried twenty-one rounds of pistol ammunition, one bandolier with sixty rounds of rifle ammunition, and one thousand rounds of automatic rifle ammunition), their other needs were grossly neglected. Myers' moving narrative of the unforeseen emergencies he encountered and the ingenuity he employed in meeting the needs of both his men and their mounts provides graphic insight into life with the United States Cavalry in the 1920s. Myers explained that he was

down there about three weeks and it may be of interest to you to know how poorly prepared we were for taking the field at that time. For example, one of my men was injured by diving in the river where the water was too shallow, and cut his head open considerably. We had no medical kits along with us . . . except the little first-aid pouches . . . and all that contained was a bandage.
   Porvenir consisted of a cotton gin and a store run by Mrs. John [Maria] Daniels. She gave me a needle and some thread and I was

---

4. Special Orders no. 50, April 5, 1929, United States Army Continental Commands, 1920–42, RG 394.
5. Special Orders no. 44, March 26, 1929, RG 394.

able to sew this man's head up with that. I guess it took about a dozen or fifteen stitches. . . . But we had no iodine, no antiseptic, no nothing, and she didn't have any either. So she gave me some kerosene, and the only treatment he got until his head healed up was a liberal dousing of kerosene and the bandaging of his head every day.

Another thing shows how poorly equipped we were and how we suffered from pre-planning [or lack of preplanning] for such an occasion. I was sent down there with thirty men and thirty-two horses and no forage. I was told to graze 'em, but horses can't work like that, ride thirty miles every day without grain. . . . Nearby was a cotton gin run by a man named Hawkeye Townsend. And he gave us from his farm a little bit of kafir corn and also some cotton seed, which the horses didn't like, but when they got hungry enough they would eat. I had to make a promise that some day he would get paid for that, but I was never sure.

Another thing, our food got pretty boring because we were on iron rations [dried and canned foods], and going three weeks on that stuff was pretty bad. So I let the men from time to time shoot a deer. That augmented our ration, although the season wasn't open. . . . But in the process of doing so, we had to expend some ammunition. . . . And I had to account for every single cartridge. (Interview, 1980)

On Saturday, April 6, a mounted detachment from "A" Troop relieved Myers' platoon.

Mexican deserters and renegade soldiers from the fighting adjacent to the Mexican border continued to pose a threat to the villages along the Texas side of the Rio Grande. Shortly after Myers returned from Porvenir, 1st Lt. Kenneth B. Hoge's "B" Troop received orders to proceed with all haste to Redford, Texas, a small village about fifteen miles below Presidio. The post commander had received an emergency report that the post office at Redford had been raided by Mexican bandits. Hoge's troop left Camp Marfa immediately. With the rough terrain over which they had to march, they did not reach Redford until late the following day. Myers recalled that Hoge's assignment also ended in disappointment; "when he arrived the bandits had disappeared" (interview, 1980).

When Hoge's platoon returned to the post he was relieved by Myers' unit on emergency orders. The latter's instructions stated that he and his platoon, with two automatic rifles, were to be "in immediate readiness to leave Camp Marfa . . . prepared to go into action against marauding bands reported operating in this sector, or to respond to any other emergency call."[6] The orders' duration time subsequently expired and they were never executed. With Escobar's armies maneuvering along the international border, however, other assignments awaited the troops of the First Cavalry Regiment.

At the same time "B" Troop monitored the fighting across the Rio Grande from Presidio, Gen. George Van Horn Moseley, commander of the United States Seventh Cavalry Regiment stationed at Fort Bliss, watched with uneasiness the developments in adjacent Ciudad Juárez. This was the largest city on Mexico's northern border and that country's most important port of entry. With a rebel victory contingent upon controlling access routes for foreign imports, especially those adjacent to the United States, it was inevitable that the spotlight of battle would ultimately focus on El Paso's sister city across the Rio Grande.

When the revolution began, Gen. Manuel J. Limón commanded the customary small peacetime garrison at Juárez. With the growing threat of a federal loss of this critical port of entry, Gen. Matías Ramos was transferred from Mexico City to take charge of the city's defense. Ramos' arrival scarcely altered a hopeless defense posture. "With most of that part of Mexico between Juárez and Mexico City occupied by the rebels," Cullinane explained, "Juárez was isolated from federal aid" (Cullinane, 1958: 101–102). Three days after Escobar attacked Monterrey, Gen. Marcelino Murrieta's rebel army was at the outskirts of the city, cutting off any hope of reinforcing the Juárez garrison. With almost certain defeat awaiting only Murrieta's attack, General Moseley maintained constant communication with the beleaguered Ramos. According to one observer, the general was taking "every precaution to prevent incidents which might violate United States territory or international law" (ibid., p. 102). Moseley, however, feared a repeat of Pancho Villa's 1910 attack on Juárez, when stray rebel bullets killed several Americans in El Paso. The general ordered out the Seventh Cavalry in full pack and field equipment, positioning two French seventy-five cannon "near the international bridge . . . as a word of warning from the United States Government" (*San Antonio Express*, March 8, 1929).

Moseley apparently had acted independently in the matter, failing to consult with the United States War Department. The placement of cannon at the international bridge caught the immediate attention of various government officials in Washington. State Department objections ultimately reached the president. In a hastily called night confer-

---

6. "Designation of Platoon for Emergency Call," Rush to Myers, April 13, 1929. Copy in Myers' possession.

ence, President Hoover summoned several cabinet officials and Army Chief of Staff Maj. Gen. Charles P. Summerall to the White House to discuss Moseley's unauthorized actions. The president, however, pursued the matter no further than Moseley's objectionable act of aiming cannon in the direction of Mexico. When shown the Washington dispatch the following morning, Moseley, who overnight had become a local hero for his actions, stated: "I have nothing to say about this reported criticism of my actions except that I am doing everything possible to assure protection to American lives and interests in case of severe fighting" (*El Paso Herald*, March 9, 1929). The fighting that subsequently erupted adjacent to downtown El Paso was indeed severe, vindicating Moseley's precautionary action.

General Murrieta launched his attack on Juárez during the early morning hours of March 8, 1929. As the rebels and federals clashed in full view of the international bridge, thousands of El Paso spectators, seemingly oblivious of the danger, clamored for preferred locations to observe a revolution in progress. "Bullets crashed into buildings several times," one journalist wrote, "but there was a very decided tendency on the part of the spectators to risk being shot rather than to miss what was going on." Amid this carnival atmosphere citizens could be seen standing on building tops and one observer said that Scenic Drive on Franklin Mountain bristled with binoculars ("A Ringside Seat," 1929: 46).

The thirteen-story El Paso National Bank Building in the downtown business district received so much rifle fire that the personnel occupying the upper stories were removed to safer areas. United States Senator Ralph W. Yarborough, then a young attorney practicing in El Paso, occupied an office on the eleventh floor of that building and witnessed much of the fighting across the river. He lived, however, in an apartment east of the downtown area, a section where stray bullets struck some of the residences, causing the city's only casualties. Yarborough remembers that he was awakened during the early morning hours of March 8 by gunfire from across the river. He arose early that morning and headed for his downtown office. His account of the events of that day follows:

I found out on the way to the office . . . that a revolution had started. I was very much interested and when I got to the office I headed for the south window. They said, "Get away from that side of the building, a few stray shots have already hit the building!" Some broke out windows in our offices . . . but [that was done] before I got there. (Interview, November 25, 1980)

Yarborough's eleventh-story vantage point afforded him an excellent overview of the battle's progress. Accepting the danger incident to his location by a south window, the former West Point appointee followed the fighting with a military scholar's interest.

> The thing that impressed me most was I could see somebody that looked like he was taking part in the fighting. He had on white clothes, you know, like peons' work clothes. They would be on tops of buildings over there, and some apparently had black powder because you could see powder puffs when they would fire. But mostly it was just scattered firing. Just desultory fighting—casual firing—after a battle is over. (Ibid.)

During the course of the battle it appeared to the El Paso observers that General Murrieta was staging the fight especially for their benefit. As his rebel troops swept through the city, they drove the defending federal troops to the river bank where they remained most of the day. With most of the fighting concentrated adjacent to the international boundary, General Moseley grew increasingly apprehensive for the safety of those people living along the north bank of the Rio Grande. By midday, when it became apparent the battle was a rebel victory, Moseley crossed the international bridge to confer with the federal commander, General Ramos. He then requested permission to cross the federal lines to warn the rebel commander against firing across the river. Ramos, however, admitted to Moseley that the federal cause was hopeless and they

> would be willing to quit if the rebels would cease firing and consent either to grant guarantees of safety or permit the Federals to be interned on the American side. General Moseley then walked to the rebel headquarters in the Mesa Hotel, where he met General [Miguel] Valles, who accompanied him to the Custom[s] House, which the rebels took after a brief skirmish late in the morning. Here the details of capitulation were discussed and a truce called. ("A Ringside Seat," 1929: 46–47)

Later that afternoon remnants of the federal garrison surrendered to Moseley, who interned more than three hundred officers and men along with their wives and children at Fort Bliss.

Although the victories along the Arizona and Texas borders enhanced the rebel image in Mexico, they also cast a lengthening shadow of fear in the United States, a fear that further strengthened that country's resolve to render immediate aid to the Mexican government. The day following the surrender of Juárez, the State Department approved a request to supply Mexico with war surplus rifles and ammunition and reaffirmed the previous agreement to allow "the Mexican government to import privately made airplanes from the United States." With this official approval "went an embargo on unauthorized shipment of planes into Mexico," presumably "to prevent the purchase of United States built aircraft by the rebels." In order to step up the air war against the Escobar forces, the Mexican government was reportedly "offering American aviators $250 a day and that federal troops were preparing a landing field at San Luis, on the Sonoran border capable of accommodating 15 to 20 planes. This was taken to mean an aerial attack [was] in preparation on the rebel stronghold of Nogales" (*El Paso Herald*, March 9, 1929).

While no one probably grasped the import of this announcement, it conveyed prophetic overtones. The airplane was approaching the threshold of military acceptance in strategic planning, and within the coming weeks it would successfully prove its combat efficiency for the first time on the North American continent. The struggle for the ports of entry would add a new dimension to revolutionary warfare. Fledgling wings bearing traditional tools of war would soon cast their menacing shadows all along the international border.

"... he dubbed the group
the Yankee Doodle Escadrille"

CHAPTER 3

Through deception, surprise, and attacking positions of inferior strength, the rebels achieved most of their initial victories. But Escobar soon encountered one aspect of civil warfare he obviously had failed to anticipate: combat aircraft. His tactical disadvantage without air power became glaringly apparent during his occupation of Torreón.

At 10:30 on the morning of March 16, 1929, two federal airplanes appeared over Torreón. One, a red-winged monoplane (type unidentified), dropped six or eight small bombs on the rebel troop trains parked in the railroad yards, and then bombed the military base located about one mile east of the city. The second aircraft, a silver-colored biplane (apparently a Vought "Corsair"), strafed the rebel troops with machine gun fire, which the rebels returned without causing any apparent damage to the plane. The two airplanes repeated the attacks several times the following day. According to James C. Powell, Jr., United States vice-consul at Torreón, the rebels sustained no casualties, while "some twenty civilians were [reported] killed or wounded by the exploding bombs. . . . The actual damage . . . was insignificant as compared to the moral [sic] effect resulting therefrom."[1] One morale casualty of these aerial attacks undoubtedly was General Escobar; one week later he was organizing his own air force.

Escobar scheduled his initial meeting on the night of March 23 in Juárez. This was an appropriate location; the border city gave the rebel general immediate but illegal access to United States pilots and aircraft. Seated around a table in the dingy Juárez telegraph office,

---

1. The rebel forces had two aircraft based at Torreón, but they did not appear during the federal attack. Powell to secretary of state, March 18, 1929, RG 59.

Escobar made his appeal to a group of five nondescript American pilots and soldiers of fortune. A common interest drew them to this clandestine setting: they hoped to reap huge dividends from the rebel coffers. Fortunately for recorded history, Tom Mahoney, city editor of the *El Paso Post*, attended the meeting. A half-century later he could not recall why he was accorded this honor. Mexican generals' traditional love for publicity, especially having their pictures taken, is a fair assumption.

Speaking partly in English and partly in Spanish, Escobar offered each pilot joining the rebels $1,000 a week, his expenses, plus a preference in granting concessions when the rebels won the war. Initially, the Americans were noncommittal; they came to the meeting expecting considerably more. Previous newspaper dispatches reported weekly salaries of $1,250 plus a bonus of $600 for each mission flown. After some hesitancy, however, all agreed to embrace the rebel cause, whereupon Gen. Gustavo Salinas, commander designate of the rebel air force, passed the muster roll for their signatures. Before adjourning to Tom Hanlon's bar to celebrate the occasion, Mahoney appropriately dubbed the hastily organized group the "Yankee Doodle Escadrille" (Mahoney, 1929: 13–14). The name stuck.[2]

The nucleus of the rebel air force contained a wide range of flying skills and expertise. Surprisingly, these mercenaries would evolve as some of Escobar's more loyal and enduring followers. The group included Art J. Smith, St. Louis, Missouri; Pete Stanley (he also called himself Stanley Thompson), Harrison, Arkansas; Jack O'Brien, San Francisco, California; Pat Murphy, who claimed no home; and Richard H. Polk from Nashville, Tennessee. (Polk was one of two professional revolution followers who joined the rebel group. He learned to fly during World War I, claimed four German victories, and, following the armistice, made a career of "revolution hopping." Prior to his arrival in Mexico, he claimed to have been employed by Gen. Augusto César Sandino, whose revolutionaries were creating problems for the United States Marines in Nicaragua.)[3] Phil Mohun, who gave his home town as Kansas City, Missouri, also elected to switch sides

2. Mercenary combat aviators engaged by foreign government follow closely the history of aviation. N. de Sakoff, a Russian hired by the Greeks in 1913 to bomb a Turkish stronghold in a French-built plane, was one of "the first in a long line of aerial soldiers of fortune who choose to make a living fighting somebody else's war" (Seagrave, 1981: 15–16).

3. In 1928, a predecessor Vought "Corsair," the O2U, flown by United States Marine Corps pilots, had also seen action in Nicaragua, successfully attacking positions held by the rebel Sandino (Moran, 1978: 41; Sherrod, 1952: introduction).

during battle. He joined the rebel group after abandoning the federal air force in the vicinity of Monterrey.

The career of General Salinas, who commanded this reckless group, provides an insight into the fluctuating fortunes of rebel allegiances in Mexico's tumultuous revolutionary environment. Unfortunately, Salinas was poor at picking winners and twice allied himself with losing causes. Mahoney remembers Salinas as a "cultured American educated Mexican officer who was graduated from Manlius Military Academy and learned to fly at Long Island. He received U.S. Pilot's license No. 172, one of the lowest numbers in existence [in 1929]" (Mahoney, 1929: 13–14). Returning to Mexico, he joined the ill-fated Gómez-Serrano revolution in 1927 and was subsequently exiled by the Calles-Carranza government. On learning of the rebellion led by Escobar, he eagerly offered his services to the new revolt, receiving a general's rank in what would become another hopeless endeavor.[4]

Mohun emerged as a key member of the Yankee Doodle Escadrille. He symbolizes the general attitude, expectations, and responsibilities of those itinerant American pilots who cast their lot with Escobar. They expressed no social or political philosophies; their prime motivations were money and adventure and they received a goodly measure of both. Mohun admits that they fought some battles, bombed some cities, wrecked some trains, and "a lot of men were killed. But for us . . . it wasn't a war—it was just the grand old revolutionary game of gyp-the-leader." He recalled he was "loafing in New York, waiting for something that looked like easy money to turn up, when the Mexican revolution started. . . . So I went to Washington and offered myself to the Mexican embassy" (Mohun, 1934: 32). This led to an assignment with Gen. Juan Almazán's federal army, flying over Torreón once a day and dropping hand grenades on the rebels holding the town. When Mohun learned of the attractive salaries Escobar was paying American pilots, his position with the federal army suddenly lost its appeal. "I was getting more and more fed up with the federals. . . . [So] I decided I'd better change sides quick," he explained (ibid., p. 34). Mohun purposely avoided giving notice to his current employers, as he journeyed to his new assignment in a purloined federal airplane.

As the United States arms embargo forbade the sale of airplanes to the Mexican rebels, alternate methods of acquisition had to be de-

---

4. Despite his rebel affiliations, Gustavo Salinas played a major role in the development of Mexican civil aviation. As director of aeronautics in 1928, he was instrumental in inaugurating the Compañía México de Aviación air mail route between Mexico City and Tampico (see Villela, 1964, and Newton, 1978).

Maj. R. L. Andrews, commander of the Mexican federal air force in Sonora, bombed an aircraft being delivered to the rebels at the Agua Prieta airport. Andrews (wearing a checked shirt) surveys the results. Courtesy R. L. Andrews.

vised. Escobar's salvation soon evolved in the person of Phil Mohun; the initial rebel fleet of airplanes was purchased in his name as an American citizen and flown to El Paso. "Then, when the chance came," Mohun wrote, "one of us rebel pilots would hop them across the border. At one time the United States Department of Commerce records listed me as the owner of nine planes, though I couldn't have bought a box kite" (ibid.).

United States dealers quickly recognized the rebels' desperate need for aircraft and also joined the old revolutionary game of gyp-the-leader. The purchase of an American Eagle two-place biplane powered with an OX-5 engine, worth about $1,000 at Phoenix, Arizona, is a case in point. "The rebels agreed to buy the crate for $3,750," Mohun explained, "if it was delivered in Mexico; so I was sent to Phoenix to fly the plane to Juárez" (ibid.). While good judgment may have indicated a clandestine entry into Mexico under these circumstances, Mohun chose to flaunt his achievement to the border authorities. He "brought . . . [the American Eagle biplane] from the El Paso airport so openly that he said he flew low over the international bridge to drop a quarter toll," Mahoney claimed facetiously. However, "this bravado [buzzing the international bridge] later cost him several weeks in jail" (Mahoney, 1929: 14).

Relying mainly on Mohun's name, skill, and daring, the rebels quickly assembled a bizarre collection of so-called fighting equipment drawn from a variety of sources, all illegal. At maximun strength the rebel air force consisted of only ten airplanes, all civilian types (except a captured "Corsair") manufactured in the United States. With the one exception, none were designed or equipped for combat assignment. The inventory consisted of a Cruisair monoplane, an American Eagle biplane, a Vought "Corsair" biplane (captured from the Mexican federal air service sans machine guns), a Thunderbird biplane, a Fairchild monoplane, a Standard biplane, a Ryan Brougham monoplane, a Travel Air biplane, and two Stinson "Detroiter" monoplanes. The Travel Air, originally flown for the federals by another mercenary, "Buzz" Morrison, was later captured by the rebels when an empty gasoline tank forced it down in Sonora. The two Stinson cabin planes belonged to a Mexican airmail company. Shortly after the outbreak of the rebellion the airmail pilots, Antonio Cárdenas and Arturo Jiménez, seized the two planes and joined forces with the rebels (ibid.). Escobar subsequently selected one of the Stinsons as his personal executive transport.

The armament carried by Escobar's rebel pilots was as ill-suited for air combat as were the airplanes. Lacking both machine guns and the technology to equip the aircraft for forward-firing pursuit and low-altitude strafing attacks, hand-held Thompson submachine guns and small caliber hand guns constituted their aerial firepower. And none of these could be fired forward through the propellers. Yet these mercenary fighters exhibited raw courage and "unhesitatingly attacked federal planes, 50 miles an hour faster and armed with synchronized machine guns" (ibid.).

The rebels were even less adept at aerial bombing. In addition to lacking training, the proper equipment, and administrative discipline, members of Escobar's flying circus nevertheless went forth into battle bearing even greater handicaps. First, dependable explosive devices were always in short supply. With the exception of a few lots of bombs smuggled from the United States, the rebels were forced to rely almost entirely on the crude and unpredictable units fabricated at the Greene-Cananea plant in Sonora.

Their aircraft also lacked the under-wing military-type bomb racks necessary to achieve any degree of bombing precision. Instead, the rebel observer-bombardiers carried their bomb loads in their laps during a mission. Approaching the target area, they carefully removed a safety screw on the bomb before dropping it over the side of the airplane or through a hole cut in the floor of the rear cockpit. Mahoney explained that "General Fausto Topete, the rebel leader in Sonora, paid the flyers there according to the number of screws they brought back" (ibid.).

Near the end of the conflict the rebels were forced to turn to even less sophisticated explosive devices: handmade bombs. A United States Army Air Corps pilot assigned to the Naco sector reported the rebels fabricated the bomb casing from a short length of ten-inch cast-iron water pipe, which they filled with whatever shrapnel material was available—rocks, iron bolts, or metal fragments. The ends were then sealed with melted lead or other restraining material. The rebels then drilled a small hole through the pipe, poured in a charge of powder, and inserted a length of flexible fuse. The use of this device, also carried in the bombardier's lap, meant that this brave man had to smoke a cigar during the bomb run. Col. Elmer P. Rose explained:

When the pilot signaled it was time to drop the bomb, he [the bombardier] had to start puffing on his cigar to light the fuse and drop this water pipe out. Well, I never saw the trajectory of it, but some of the people said that it was wonderful to behold, because it went in every direction under the sun except where you would expect it to go. (Interview, January 28, 1980)

Although the Yankee Doodle Escadrille gave the Escobar forces added mobility as well as a surveillance and reconnaissance capacity, tactically it seldom constituted more than a general annoyance to the federal forces. The rebel air force did, however, score some isolated successes. This accorded the group combat status, establishing it, at least, as a visual threat to be both attacked and avoided by the federals. The rebel air force served in most combat theaters of the revolution. News reports, however, indicate it achieved its major successes in the Sonora sector during the battles for the ports of entry. For example, a March 31 item states that "bombs dropped from an altitude of 4,000 feet by a rebel aviator late this afternoon scored two almost direct hits in the Naco [Sonora] trenches and killed at least two Federal soldiers" (*San Antonio Express*, April 1, 1929). There would be more rebel bombings along the Sonora-Arizona border, some displaying far less accuracy. Such irresponsible action would not only hasten the demise of the Yankee Doodle Escadrille but the revolution as well.

Events that followed the March 20 meeting in the Juárez telegraph office gave the rebel leader cause for concern. In retrospect, Escobar's decision to organize an air force appears as a desperate move to salvage the rebel effort. After abandoning Saltillo and Torreón to Calles' advancing federal army, news of the fate of his coconspirator Gen. Jesús M. Aguirre reached Escobar in Juárez. After he deserted his Veracruz post, pursuing federals captured and tried Aguirre on March 20 and executed him the following day. With the Veracruz rebellion quieted, the government forces controlled the Gulf of Mexico, the southwest, and Monterrey. Calles could now turn his attention to the north and west, Escobar's two remaining strongholds. This campaign would again push the conflict nearer to the Texas border.

Escobar undoubtedly recognized by the end of March that the high tide of revolutionary fervor would soon spend itself; his future conduct of the war would place victory or defeat in delicate balance. He was also aware that his leadership role was in deep jeopardy. Some of the rebel leaders challenged his conduct of the campaign, as he "seemed to be capable only of retreating, destroying railroads, sacking banks, and making newspaper statements. Consequently, the post of Supreme Commander was offered to General Marcelo Caraveo," who rejected the appointment (Valenzuela, 1975: 47). Escobar, nevertheless, seemed to ignore the criticism of his peers and continued to pursue his goal at a fanatical pace, hoping to avoid further contact with the federals until he could regroup his forces at some other location.

After giving up Escalón to federals, Escobar continued his northward retreat and assembled about eight thousand men at Jiménez, where he decided to engage the federals for control of this important southern Chihuahuan trading center. In this bloody encounter, which marked the turning point in the revolt, both federal and rebel air units met in the skies over Jiménez. The federals' recently acquired Vought "Corsairs" helped deliver the winning punch and further hastened Escobar's political and military demise. General Calles first called on his air force as he advanced toward Jiménez. While he marched across the Bolsón de Mapimí, the "Corsair" pilots scouted the route for watering places for Calles' five thousand horses and nine thousand-man army and, according to Dulles, "reporting fully on the location and estimated size of enemy forces" as he maneuvered for his assault on Escobar's rebels (Dulles, 1961: 449).

The five-day encounter at Jiménez began in Corralitos at 9:00 A.M. on March 30 and ended at Reforma Station (La Reforma), about twelve miles north of Jiménez, at 2:00 P.M. on April 3. By the end of the first day's fighting, half of the city was in federal hands. General Calles informed the government later that night that the "Federal artillery had resumed bombardment on Jiménez after a lull in the fighting earlier in the day. At 5:00 P.M. he said shelling of the city was begun again and a windmill which had served as a rebel machine gun tower was destroyed" (*San Antonio Express*, April 2, 1929).

The federal sweep continued the next day with Calles' forces scoring a resounding victory. Newspaper headlines described the rebel carnage left in the battle's wake: "Rebel Force in Disorderly Retreat After Two-Day Resistance; Bridges Burning; Dead Horses and Men Strew Streets of Town, Air Observer Says." The federal air force's low-level strafing and bombing attacks inflicted death and destruction on a virtually defenseless rebel army. Lt. Col. Gustavo León, a federal pilot, "reported that the rebels were fleeing from Jiménez and that he bombed them and fired at them with his machine guns. The bombs fell 'exactly on the enemy columns' he said" (ibid., April 3, 1929). A battlefield observer described the encounter as "true butchery," while General Calles called it a "decisive defeat for the rebels and the greatest battle in Mexico's revolutionary history" (ibid., April 4, 1929).[5]

While the strength of the federal air force at Jiménez is not known, the State Department reported that seven rebel aircraft, based at Chihuahua, engaged in the conflict (April 3, 1929, RG 59). The aerial encounters, however, proved less decisive than the "Corsair" attacks on the rebel ground forces. During the course of the Jiménez campaign both sides lost an aircraft to enemy ground fire, but in the end the rebels could claim a victory of sorts. The federals suffered the first loss. Prior to the battle, Lt. Col. Roberto Fierro, commander of the federal air corps, received orders to fly a bombing–propaganda tossing mission over the rebel lines at Jiménez. Sighting the federal plane approaching their installation, Phil Mohun and Richard Polk took off in their vintage aircraft to challenge the enemy "Corsair." Simultaneously with their air attack, rebel ground installations opened fire on Fierro's plane, and "within a few minutes a bullet hit the carburetor of the Vought." (Mohun claimed the victory.) In spite of the engine failure, Fierro was able to land the plane unhurt, remove the machine gun, and, with his observer, a Lieutenant Valle, escape into the mountains ahead of pursuing rebel horsemen. "The rebels speedily repaired Fierro's plane," Mahoney adds, "and put it into use" (Mahoney, 1929: 15).

Suffering from hunger and thirst while hiding in the mountains, Fierro eventually made his way back to the federal encampment. As-

---

5. Calles exaggerated. On April 13–15, 1915, when Pancho Villa led a charge of about twenty-five thousand troops against Gen. Alvaro Obregón's fifteen thousand defenders at Celaya, four thousand Villistas were killed and six thousand taken prisoner. Dulles pronounced this a demonstration of raw courage versus Obregón's more scientific warfare. "With the exception of the Civil War in the United States, the American continent had seen no encounter as colossal or sanguinary as was this" (Dulles, 1961: 12).

signed another aircraft, he again took to the air, this time delivering the rebels a crushing blow with a critically placed bomb. Flying a reconnaissance mission over the battle lines, the young colonel watched with concern as it appeared the federal forces were being lured into a rebel trap. What Fierro observed was an exercise in basic military strategy. Escobar's plan was to allow the federals to beat back his flanks and surround Gen. Miguel Valle's forces in the trenches. With the federal forces committed, the rebels would counterattack from the outside while Valle attacked from his surrounded trenches, thus catching the federals in the cross fire.

The rebels appeared to be executing the strategy according to plan, and had it not been for Fierro's astute aerial reconnaissance history would undoubtedly record Jiménez a rebel victory. Circling the battle area, he sighted a string of boxcars behind the rebel lines, one of which he discovered later was loaded with dynamite used by Escobar's forces to destroy railroad bridges. As Fierro moved into position to make a bombing run on the unprotected train, Mohun and Polk took off in pursuit of him. Their effort went for naught as their underpowered aircraft were no match for the faster "Corsair." On the first two runs Fierro scored near misses on the rebel supply train. On the third pass, Mahoney reported, the bomb

> struck the car of dynamite and there was a tremendous explosion. The rebels, five miles to the south, thought that they had been surrounded. . . . General Escobar's regiments, instead of continuing to the aid of General Valle, broke ranks and ran. As they fled, Almazán's cavalrymen cut them down. Of the 800 men with Valle, only 2 escaped alive. (Ibid., p. 72)[6]

During the Jiménez encounter the rebel air force also emerged unquestionably the loser. In one engagement Polk sustained a bullet wound in the face and severely damaged his aircraft while attempting an emergency landing. Mohun likewise suffered from the greater firepower of the federal "Corsairs." After Mohun's American Eagle biplane was riddled by machine gun fire at 1,500 feet, he rode the crippled

---

6. In 1921, Roberto Fierro first distinguished himself as a skilled aviator while attending the government flying school in Mexico City. He received international honors for his goodwill flights to Cuba and Central America, and was later appointed head of the Mexican federal civilian aviation department for his role in defeating the Escobar rebels. He later clipped more than ten hours from the New York to Mexico City nonstop record set by Col. Charles A. Lindbergh (see Fierro Villalobos, 1964; Villela, 1964; and Tom Mahoney manuscript on Roberto Fierro in my personal files).

plane down to a crash landing. (He admitted that this was not an act of heroism as "we had only one parachute and we took turns wearing it." A later attempt to procure a second parachute ended in failure. Mohun explained that "a pilot was given $400 and told to purchase the chute in El Paso. He didn't come back" [Mohun, 1934: 34].)

All and all, it was a bad day for the rebel air force. In addition to Mohun's American Eagle casualty, the "Corsair" captured from the federals was also severely damaged by machine gun fire. Unable to make the repairs in the field, the rebels later loaded the three airplanes on a rebel troop train retreating toward Juárez. Ambassador Morrow, who closely monitored the Jiménez fighting, telegraphed the secretary of state on the morning of April 3 that "a message received from General Calles at 3 o'clock this morning reports rebels leaving Jiminez [*sic*], some by train, some by truck and some on foot. Several railroad bridges north of Jiminez have been destroyed by Federals. Information not yet definite as to possibility of intercepting retreat" (April 3, 1929, RG 59).

Escobar survived the Jiménez disaster, leaving in his wake another tragic testament to the cost of rebellion: 1,136 men killed and 2,058 wounded or taken prisoner (*Foreign Relations*, 1943:3:423). Not included in these figures were the large numbers of deserters and escapees. The latter found no solace in defeat as federal cavalry detachments pursued the "rebels who escaped being cut to pieces yesterday. . . . Loyal contingents were . . . [rounding up] the dispersed insurgent infantry, small groups of which were roving about the plains north of La Reforma seeking a means of escape" (*San Antonio Express*, April 5, 1929). Many of these renegades did elude the federal cavalry, fleeing northward. Eventually some would reach the Big Bend country to spread fear and panic among the settlements along both sides of the Rio Grande.

Five troop trains bore the remnants of Escobar's command—fewer than two thousand men and their mounts—in rapid retreat from La Reforma. Following a brief stop in Chihuahua, they continued on to Juárez, arriving there on April 6. Escobar escaped in one of his airplanes, flying ahead to Juárez to await the arrival of his troops. His first order of business was to confiscate all the city's taxi cabs—his purpose not disclosed—and order them driven to the railroad yards. "By nightfall," Cullinane wrote, "87 taxis were in the yards and all were loaded on a sixth train." With this additional cargo the rebel convoy continued the retreat to Sonora, proceeding over the Mexican Northwestern Railway to Casa Grandes, and thence via Pulpito Pass to Agua Prieta, arriving there on April 13. Escobar followed his troops, probably by air, but before leaving Juárez he proclaimed himself the provisional president of Mexico (Cullinane, 1958: 105).

With the rebels' departure from Juárez, the evacuation of that city was virtually complete. Only a token force remained to hold the city in the face of the pursuing federals. Recognizing the futility of his position, the former conqueror of Juárez appealed for asylum in the United States. On the morning of April 12, District Director Wilmoth telegraphed his request to the Immigration Bureau in Washington.

Upon receiving report of a small body of federal troops marching from Guadalupe to Juárez, General Murrieta decided to retire rather than have further bloodshed, and upon his application for asrlum [*sic*] was admitted this port as a refugee, being assessed temporary deposit as a visitor. He was paroled upon promise to remain El Paso and not engage in revolutionary activities. He is without funds and requests permission to proceed Los Angeles where he claims he can earn livelyhood [*sic*] for self and family either by teaching or as a watchmaker. (April 12, 1929, RG 85; punctuation added)

The advancing federal army entered the city, and by April 9 Juárez was again under federal control.

The defeat at Jiménez and the ensuing retreat to Sonora mark the turning point in the rebellion. And while Escobar may have designated himself Mexico's provisional president, the proclamation was merely political rhetoric. The revolutionary movement had lost its momentum as well as its unity. Coordinated command no longer existed and desertions and defections were gradually decimating the rebel army. The remaining pockets of rebel strength stretched along the Sonora-Arizona border between Agua Prieta and Nogales, where Escobar planned to join his forces with those of coconspirator Gov. Fausto Topete. Prospects of altering the course of the conflict, however, were not promising. Nogales, Sonora, remained the rebels' last major stronghold, yet the town was completely surrounded by the federals on the Mexican side of the border. On the north stood the international boundary fence secured by the United States cavalry from Fort Huachuca. Escobar's strategy, however, remained unchanged: gain control of the ports of entry and hope that outside aid could be drawn through these corridors from the north. Here they would make one last stand to salvage the rebel cause.

Desperation bred desperation, and in the rebels' effort to dislodge the federals, they violated United States territory. Repercussions were inevitable.

"... two more bombs have fallen just within American territory"

CHAPTER 4

Although troops of the Eighth Corps Area at Fort Sam Houston had been alerted for border duty since the beginning of the rebellion, the continuing struggle for control of the Arizona ports of entry eventually triggered full-scale mobilization at the San Antonio base. Meanwhile, army intelligence continued to monitor and evaluate Mexican troop movements along the international border. "Every move Federal and Rebel Mexican forces make is watched closely," reported the *San Antonio Express*, "and as the scene of activities moves westward there is a westward movement of troops on the American side of the border." The military's immediate objective was to insure that combating units remained on Mexican soil and the lives and property of American citizens were not endangered. The report warned that "should any desperate rebel forces, prevented from escaping to the south or west by the Mexican Federal forces, decide to escape through American territory, they will be met at the border by American soldiers" (*San Antonio Express*, April 10, 1929).

Fluctuating allegiances of the Mexican military leaders, both rebel and federal, compounded further the task of determining the changing fortunes of war. This was equally confusing to the correspondents reporting the action. It was "necessary to cover the gossip-front even more diligently than the battles," explained Duncan Aikman. "The 'Who's Who' of the revolution was important, but 'Who's on Which Side To-day' was the chief issue." For example, when the rebel commander at Naco, Gen. Agustín Olachea, and his troops "renounced the rebel cause and were henceforth loyal to Federals," Gen. Fausto Topete ordered 1,500 rebel soldiers from Nogales to reinforce the

Naco garrison in an effort to retake their former position (Aikman, 1929: 243).[1]

Topete established his command post at Mina Oro, six miles south of Naco, Sonora. The Yankee Doodle Escadrille launched twice-daily bombing raids from Mina Oro against Olachea's federal installations at Naco. Two serious problems faced the American mercenaries. First, their primitive equipment and quaint methodology made precision bombing impossible, and second, to further complicate their assignment, their target was Naco, a small Mexican village separated only by a wire fence from its sister community in Arizona. Either from lack of skill or from being unable to identify their target, several rebel bombs fell wide of the mark. Immigration Inspector Walter F. Miller, who witnessed the attack, reported: "At 7:45 AM on April 2nd a rebel airplane dropped two bombs on the U.S. side of the line in Naco, Arizona, one of which bomb fell a little over a hundred feet from this office and broke one window. . . . One U.S. citizen was wounded by a flying fragment of the bomb and was taken to this office for treatment" (May 17, 1929, RG 85).

Just prior to this attack, Maj. Gen. William Lassiter, commander of the Eighth Corps Area, had ordered Brig. Gen. Frank S. Cocheu to Fort Huachuca, Arizona, to assume command of all military forces along the Mexican border. Immediately following the bombing, Cocheu, accompanied by the United States vice-consul at Agua Prieta, called on General Topete at his headquarters at Mina Oro. He explained that

> the United States government would not countenance the injury of American citizens and damaging of American property by either rebel or Federal forces. General Topete expressed deep regret that bombs had fallen on American soil, stating that it had been entirely accidental and promised that it would not occur again. He instructed his commercial agent here, who was our guide, to settle all damages at once. (*Foreign Relations*, 1943:3:380)

The entire matter had a familiar ring. During a previous Mexican military encounter American citizens had also felt the sting of battle. In December 1914, United States troopers installing telephone lines

---

1. General Olachea was probably the victim of a "$50,000 cannon ball," a term reflecting some rebel generals' well-known greed. The process was simple. Federal officials negotiated with the rebel leader for the purchase of a symbolic cannon ball. The sale usually included amnesty for the leader, sometimes reinstatement at his former rank, and the surrender of his entire command, which had usually joined the rebel forces with him.

along the Arizona border near Naco became the targets of Mexican Constitutionalist riflemen, and the incident provoked protest by the American officials at Naco. Gen. Tasker Bliss ordered three long-range batteries of the Sixth Field Artillery to take up positions adjacent to Naco and dispatched orders to the opposing leaders of the fray, Gen. Benjamin Hill and governor of Sonora José María Maytorena, to cease firing or risk an artillery attack (Mumme, 1979: 157).

And now, fifteen years later, the United States was again attempting to quell a borderland threat with diplomacy. History, and General Bliss, had demonstrated the difficulty in gaining rebel leaders' attention with mere verbal threats. Such matters would ultimately require firmer measures.

Topete, as expected, fulfilled only a portion of his verbal commitment. The injured American, identified as Harry Baker of Alliance, Ohio, was transferred to a Douglas, Arizona, hospital where he received treatment for a scalp wound and a "satisfactory settlement" negotiated by Topete's local representative. However, following a one-day moratorium to honor two fallen enemy airmen, the bombings continued.

On April 4th, a Federal airplane while bombing rebel troops some few miles south of Naco, was shot down in flames and the aviator and bomb dropper killed. Their charred bodies were brought in, under a flag of truce, by the rebels that evening and General Topete announced that out of respect to them, his aviators would not bomb Naco, that evening. (Miller, May 17, 1929, RG 85)

The two airmen were part of the Mexican federal air force detachment commanded by Maj. Rayma L. Andrews, a United States citizen under contract to the Mexican government. The federals, as did the rebels, relied on imported United States personnel to help organize and administer the combat air units. Andrews, an aviation pioneer, had a long association with combat aviation as well as with military leaders in Mexico and Latin America. After teaching himself to fly in 1910 in a home-built Curtiss pusher, he joined Glenn Curtiss the following year on an airplane-selling junket to South America. In Ecuador, Andrews demonstrated the airplane's military potential by dropping explosives from the Curtiss biplane. Andrews explained the procedure: first, the pilot smoked a cigar, which he touched to the frayed end of the fuse; second, as he watched the fuse burn toward the box of explosives attached to the airframe, he clipped the wire securing the box of dynamite with a pair of pliers lashed to his wrist; and third, he kicked the box of dynamite free of the aircraft. "It usually exploded either in the air or on the ground," Andrews added. "It sure impressed those generals watching the demonstration. We sure sold a lot of airplanes that way" (interview, February 18, 1981).[2]

Andrews continued his foreign military career during World War I, flying British SE-5 pursuit planes for the British Royal Air Force. His association with Mexico began in 1919 when he and another American aviation pioneer, Eddie Stinson, flew to Chihuahua on another airplane-selling venture. Andrews knew the governor of Chihuahua, Gen. Ignacio Enríquez, an early aviation enthusiast, and they sold him a used Italian Ansaldo C-300 six-place cabin biplane.

During the rebellion Enríquez remained loyal to the Portes Gil government, serving as commander of the Fifth Military District of Mexico. When it appeared that the revolution would be won or lost in the northwestern territories, Enríquez asked Andrews to come to Chihuahua and organize an air force for the defense of that region. The offer included an attractive financial arrangement, plus the rank of major in the Mexican army. Resigning his second lieutenant's commission in the United States Army Air Corps Organized Reserve, Andrews accepted the assignment. His reason for joining the federals: "Partly for the money, but mostly the adventure" (ibid.).

Andrews served the federal army both as a pilot and an administrative officer. Procuring equipment and supplies was his initial responsibility as well as his source of income. In purchasing airplanes, spare parts, parachutes, and other accessories, Andrews determined the total cost and added his commission. The Mexican government then deposited the specified amount in the El Paso National Bank, on which he was authorized to issue drafts. "What was left over was mine," Andrews explained. He purchased a wide range of American-built commercial-type biplanes that included Travel Airs, Eaglerocks, and "about seven Stearman biplanes. I usually flew the Travel Airs." But late in the conflict, the "Mexican government sent up some of the 'Corsairs' to Naco, and I flew them from time to time" (interview, July 2, 1981).

From the federal airfield located just south of Naco, Sonora, Andrews flew bombing missions against the rebel ground positions surrounding

---

2. Andrews' further claim that he was the first person to drop explosives from an airplane is debatable. According to Maj. Gen. James E. Fechet, former chief of the United States Army Air Corps, "Bombs were first thrown (not dropped) by the Italians in 1911 in their Tripoli campaign, and by the Spaniards in 1913 in Morocco. . . . The first aerial bombardment in the World War [I] was that of Paris on August 30, 1914, from a single German airplane" (*Aero Digest* [November 1932]: 18; Seagrave, 1981).

Maj. Rayma L. Andrews (right) with a Mexican colonel and two Mexican pilots wearing goggles at the federal airport at Naco on April 4, 1929. The pilots were killed the following day when rebel ground fire downed their airplane. Courtesy R. L. Andrews.

that federally held border village. Provided with an abundant supply of twenty-five-pound fragmentation bombs and one-hundred-pound demolition bombs supplied by the United States government, Andrews concentrated his effort on the rebel supply centers and troop movements.[3]

While patrolling the area south of Naco, Sonora, in April 1929, Andrews sighted a rebel cavalry unit marching toward the federal entrenchments near Naco. Upon seeing Andrews' airplane, they broke column and rushed toward the overhanging banks of a dry riverbed for protection. The concentration of men and animals made a perfect

---

3. The bombs the United States supplied the Mexican air force came from Camp Stanley, a subpost of the San Antonio Arsenal, located at the Leon Springs Military Reservation near San Antonio. Andrews earlier developed a homemade aerial bomb for the Mexican federal air force. Unlike the rebel missile, Andrews' device employed a more orthodox percussion-detonating device and used metal fins welded to one end of the bomb to control its trajectory.

target. Diving his plane at the rebels, Andrews released both one-hundred-pound bombs. The explosions discharged great columns of dust that momentarily concealed the target area. Andrews circled the area waiting for the dust to settle in order to determine the effectiveness of his attack. The grim evidence was soon apparent. "Horses and men were strewn all along that river bed," Andrews remembered. "Killed eight officers. Don't know how many [enlisted] men were killed. They buried the officers. Had my picture made by one of the graves" (ibid.).

Following the one-day moratorium for the two downed federal airmen, the Yankee Doodle Escadrille again took the offensive, resuming the daily bombing strikes on the federally held village. As before, some of the explosives missed their target, falling on the United States side of the border. Again the vice-consul at Agua Prieta expressed official concern to Secretary of State Henry L. Stimson. He reported: "Rebel aeroplanes continue to bomb Naco several times daily and are drawing in closer to the town for battle which now seems imminent. Two more bombs have fallen just within American territory without damage. General Topete has again apologized to American authorities" (*Foreign Relations*, 1943:3:384).

Topete's concern for the safety of the civilian population was undoubtedly sincere. On March 29 the rebels flew over Naco in the recently captured Travel Air biplane dropping handbills warning all noncombatants that "they try to keep themselves at a distance of more than three hundred meters from the Federal positions, and, if it is possible, better yet to completely abandon the town, in order to avoid personal tragedies." Referring specifically to the American citizens across the international boundary, Topete added: "Our soldiers, artillerymen, and aviators have very strict instructions to prevent their projectiles passing to the American side in order to avoid incurring an international conflict."[4] If there are American casualties, Topete argued, the federals, not the rebels, should be held accountable. He based his defense—suspect, to say the least—on the federals' determination to hold positions adjacent to the international boundary. In order to protect the civilians, Topete claimed further that he had invited the "Imposter Leaders" to come out and fight in open country, but because of their cowardice, they preferred

to sheild [sic] themselves in the civilian population and protect their back by means of the International Line. . . . The traitorous

---

4. Handbill translation, Topete to civilian population, March 28, 1929, RG 85.

Maj. R. L. Andrews with bombs (twenty-five and one hundred pounds) used by the Mexican federal forces during the Escobar Rebellion. This photograph was made by a little girl visiting the Naco, Sonora, base. The picture was later made into a postcard and sold in Bisbee, Arizona, drugstores. Courtesy R. L. Andrews.

leaders, Abelardo Rodríguez, Lucas González, and Augustín Olachea are managing to have as their base, foreign territory, across which to mobilize all their equipment. . . . (Ibid.)

Federal control of the Naco port of entry had become an increasingly volatile issue with the rebels. Access to this "foreign territory" committed to federal support meant supplies and reinforcements, ultimately the difference between victory and defeat. This strategic advantage the federals enjoyed would be further dramatized during the forthcoming encounter.

Topete launched this long-expected attack on Naco shortly after dawn on Saturday, April 6. Following a bombing attack by two rebel aircraft, some two thousand infantry and cavalry, armed with rifles and machine guns, spearheaded the ground assault. The Yaqui Indian infantry, which struck the federal trenches east of the village, gave the rebel army its most unique color. Following two armored tanks built on caterpillar tractors, Miller reported, "the Yaquis stoically advanced, single file, every third or fourth man beating the little drum carried by most of them, but the withering machine gun fire from the trenches soon forced them to take cover." The rebel cavalry, though less colorful, was equally ineffective during the attack. Exhibiting no apparent interest in dislodging the federal forces, their unit "devoted its time to an unsuccessful effort to drive a herd of stray horses through the barbed wire entanglements to create an opening for a charge," which they never launched (Miller, May 17, 1929, RG 85).

The rebels' tactical ineptness was further magnified by the unexpected arrival of federal reinforcements through the Naco port of entry. On the morning of April 5, Gen. George Van Horn Moseley, acting on directions from the United States War Department, had released 295 Mexican federal soldiers interned at Fort Bliss to the Mexican consul at El Paso.[5] Later that afternoon, accompanied by their families, they boarded a fourteen-car special train bound for the battle

---

5. The entourage consisted of "295 military personnel, 46 women, 33 children, 37 horses and some motor transport vehicles. Governor Phillips of Arizona strongly protested to the VIII Corps Headquarters and to Washington against what he termed 'invasion' of his state, but the movement was completed before anything could be done to stop it" (Cullinane, 1958: 103). See also the telegrams from Wilmoth to the Immigration Bureau, Washington, D.C., April 6 and April 13, 1929: "Unless absolutely necessary, please do not give publicity to governor's part in transaction as this might injure him in southern Arizona where residents have profited considerably from trade with rebels" (RG 85).

Mexican federal troops brace for rebel attack during the siege of Naco, Sonora. Machine gun in foreground appears to be aimed at attacking rebel aircraft. Courtesy National Archives.

front, arriving at Naco, Arizona, at 9:30 A.M. the following morning. According to Miller's report, they arrived under guard during the height of the battle "and the men were allowed to depart immediately to Mexico to aid in the defense of Naco." The women and children were paroled to the Mexican consul and remained in Naco, Arizona, until the fighting ceased, "when most of them joined their men in the trenches" (Miller, May 17, 1929, RG 85).

Throughout the battle rebel airplanes made repeated bombing runs on the Mexican village. Both sides sustained damage during the encounters. On an earlier mission one plane was hit twenty-six times by rifle fire and made a forced landing near Topete's headquarters at Mina Oro. The bombs they dropped, if they exploded, were effective only if they scored a direct hit, which was seldom. Miller reported they were of poor quality, constructed of case brass, loaded with dynamite and slugs, and weighing about fifty pounds each. Although the Yankee Doodle Escadrille dumped more than one hundred bombs on Naco before and during the battle, the casualty list was comparatively small. One captain, one soldier, and a civilian were killed; several soldiers and a United States citizen were wounded (ibid.).

The rebel air attacks, however, were not matters to be taken lightly. The inspector in charge of the United States Immigration Service staff reported later:

All during the battle rebel airplanes bombed the town of Naco, Mexico, and one bomb fell on the U.S. side of the line, a full block north of the Customs House. Bullets were continually fly, ing on this side of the line and this office was hit many times, although the officers remained on duty during the entire day. (Ibid.)

Nightfall brought neither victory nor defeat. The federals con, tinued to hold Naco, Sonora, and the rebels withdrew to Mina Oro, presumably to regroup for another assault. General Topete assured General Cocheu that he had ordered the withdrawal "because of the fear of causing international complications by bullets falling on Ameri, can territory." He added, however, that "he will reopen his attack at once" (*Foreign Relations*, 1943:3:188).

But all was not quiet in the wake of the rebel withdrawal. Repeated bombings of United States territory and the violation of its air space had indeed prompted a re-evaluation of the nation's defense plan along the southwestern perimeter. The orders came from the Eighth Corps Area headquarters: switch from a monitoring operation to a combat ready posture. On Sunday, April 7, the *San Antonio Light* re, ported that "for the first time since the [First] World War, Fort Sam Houston moved troops yesterday, armed, ready for action under order of war." Because of the tense border situation, the Eighth Corps Area troops "virtually slept on their arms Saturday night awaiting develop, ments in the turbulent Mexican situation. Under wartime orders is, sued by Major General William Lassiter Saturday, the entire strength of the Corps area stood ready to move at a moment's notice to protect American lives and property on the Mexican border" (*San Antonio Light*, April 7, 1929).

General Lassiter undoubtedly accepted Gen. Tasker Bliss' logic in dealing with the borderland revolutionists: cease firing or risk being shelled. During the fifteen-year interim, however, the makeup of the United States strike force had changed. Lassiter had choices unavail, able to his former colleague; one part of his response to border threat would be airborne. Two days after assigning General Cocheu to Fort Huachuca to coordinate the border defense, General Lassiter supple, mented his cavalry and infantry units with six observation planes from Dodd Field (San Antonio) and twelve attack planes from Fort Crockett (Galveston), fully armed for combat. Lassiter's orders were succinct and emphasized the importance of the mission: control the air along the border in the vicinity of Naco, Arizona (ibid., April 6, 1929). Before departing for Fort Huacheu, both squadron leaders received orders from Cocheu to attack any Mexican airplane encountered in United States air space. He explained further that "Mexican Federals at Naco, Sonora, Mexico, were advised of the instructions and were told that the inability of American airmen to distinguish between Federals and rebels in the air resulted in the ultimatum" (ibid., April 8, 1929).

Protecting vast distances along the United States–Mexican boundary threatened by revolution required speed and mobility. The United States Army Air Corps offered a measure of both. The operational speeds of most combat aircraft in the late 1920s varied from 90 to 110 miles per hour. This brought Fort Huachuca within a full day's flying time of the San Antonio area bases (depending on the wind direction) as compared to four or five days required to load, transport, and dispatch infantry and cavalry units via the railroad.

The Army Air Corps' responsibility for protecting the Arizona border fell mainly to nine Curtiss A-3 "Falcon" attack planes. These formed the Ninetieth Attack Squadron, Third Attack Group, based at Fort Crockett. This two-place biplane, identified by its upper swept-back wing (variable stagger) and needle-point propeller spinner, was powered by a Curtiss D-12 435-horsepower engine and cruised officially at 116 miles per hour. Designed primarily as a low-level attack aircraft, the A-3 carried five .30 caliber machine guns (four forward guns and one swivel gun operated by the observer-gunner-crew chief in the rear cockpit) and various combinations of light fragmentary bombs suspended beneath the lower wings.

Col. Elmer P. Rose recalled the squadron had just returned from a Saturday morning practice mission when they received orders to report to Fort Huachuca. (With the secrecy surrounding the mission, the squadron members did not know their destination until they arrived there the following day.) The date was April 6, and Topete's siege of Naco had been under way about three hours when the squadron returned to the base. As they parked the aircraft, "a messenger was waiting with a message from the group commander, who told the squadron that 'all pilots and gunners stand by their airplanes. Captain Heisen report to post operations'" (interview, January 28, 1980). Capt. Horace Heisen, a World War I United States Air Service veteran and squadron commander, soon returned, called the squadron members together, and said, "We have one hour and a half to prepare to leave for an unknown destination for an unknown period of time. And we will go armed" (ibid.).

Rose called his wife at their downtown Galveston apartment and told her to pack his bag with military clothing only, and "don't ask me any questions." When he returned to the field the ground crews had already installed auxiliary gasoline tanks and serviced the airplanes. Seeing the "belly tanks," Rose knew that a great distance lay between Fort Crockett and their destination. "We took off and headed west," Rose explained. "We didn't know where we were going, only Captain Heisen knew. I found out later that he was the only member of the squadron that had any maps" (ibid.).

Once airborne, the squadron assumed formation, and Rose, flying on Heisen's right wing, continued to watch for familiar landmarks in order to determine their position as well as speculate on their destination. The squadron passed over Kelly Field (San Antonio) and continued west. Later Rose recognized Del Rio, saw the Rio Grande off to his left, and later passed over the Air Corps auxiliary landing field west of Dryden. Just before sunset the landing field at Camp Marfa, near Marfa, Texas, came into view and Heisen signaled the squadron into landing formation.

The squadron remained overnight at the cavalry post. Early the following morning they returned to the field, warmed up the engines in the dark, turned on the running lights, and took off at first light. Continuing on their westward course, Rose remembered the squadron

> passed over El Paso and headed straight west. Out a ways in New Mexico . . . we could see that smelter down at Columbus. A little further we jogged straight south [and] . . . followed the Mexican border, watching for monuments along there. . . . We turned again and flew west. We crossed over a ridge of mountains and down into another wide valley and saw another ridge of mountains up ahead of us. We swung over and headed northwest for a while and then he [Heisen] signaled us into landing formation. (Ibid.)

This was obviously a place the young lieutenant had never seen. As he circled the field awaiting his turn to land, he noticed a big white cloth T stretched out in a big pasture as well as oil and gasoline drums provided to service the aircraft. This was not a permanent Air Corps installation, he thought. "We landed and taxied up and they [the infantry] were waiting for us," he recalled. "We tied our planes down and they took us to a national guard camp." Rose added that they still did not know where they were, but they could see an army post about a mile away. Heisen called the squadron together and explained they were at Fort Huachuca, Arizona, located southeast of Tucson about thirty miles from the Mexican border. General Cocheu had defined their mission prior to their arrival: patrol the international boundary and "attack any Mexican airplane flying over United States territory" (*San Antonio Express*, April 8, 1929).

Maj. R. L. Andrews (second from left) and another Mexican Air Force officer (fourth from left) visit with unidentified officers of the Ninetieth Attack Squadron, Third Attack Group, at Fort Huachuca, Arizona. Members of the Ninetieth Attack Squadron patrolled the United States– Mexican border in Curtiss A-3 aircraft (shown here) during the final stages of the Escobar Rebellion. Courtesy R. L. Andrews.

The following morning at first light members of the Ninetieth Attack Squadron began a patrol procedure they would follow for the next three weeks. Two two-plane flights took off together and flew south to the Mexican border. At that point they separated, one flight patrolling east, the other west. The eastbound patrol flew to a mountain range east of Columbus, New Mexico, returned to a cavalry base at Columbus, landed, and remained on alert until relieved the following morning by the next patrol. The other two-plane flight turned west at the border, carefully staying north of the international boundary, and flew to a point about fifty miles west of Nogales, Arizona. This flight then returned to the Nogales airfield where the crews remained on alert until relieved the following day (Rose, interview, January 28, 1980).

The Columbus airfield was nearest to the Naco headquarters, and they were connected by telephone. "If anything broke over there," Rose explained, "they could call us over at Columbus." But the call never came. The presence of armed United States military aircraft had

an immediate effect on rebel air activities around Naco. The squadron's mission was "to keep these renegades from coming across. Their
bombing reduced very rapidly, and finally after about a week, none at
all" (ibid.). The daily air patrols, however, continued out of Fort
Huachuca, while the cavalry units maintained constant ground surveillance along the border.

As the rebel and federal forces clashed along the Arizona border,
another battle, on the economic front, was touched off by the United
States enactment restricting trade with the rebels. For the Arizona
border merchants whose trade came largely from Mexico, this was indeed a critical issue. The delicate task of determining the political allegiance of Mexican citizens requesting border passage to purchase
supplies, therefore, fell to the United States Immigration Service. According to local reports the service expedited the border crossings in a
manner that facilitated local trade while fostering good relations with
the citizens of northern Mexico. Veteran immigration officers who
had served through twenty years of revolution noted conditions at the
ports of entry differed little from past revolts except that "for the first
time there is no hint of anti-American sentiment among the rebels"
(*Douglas* (Arizona) *Daily Dispatch*, March 17, 1929).

The airmen on temporary assignment at Fort Huachuca also noted
the lack of anti-American sentiment as well as the effect of the United
States embargo on the revolution's outcome. Lt. LeRoy Hudson, a
pilot with the Twelfth Observation Squadron who became acquainted
with both rebel and federal pilots, wrote:

> We would often at night go across the border at towns on the east,
> retaken by government troops . . . and talk with them, and on
> subsequent nights, perhaps the next, would discuss conditions
> with the rebels in the towns further west not yet retaken! It was
> quite a friendly conflict as far as we were concerned, but was real
> for the Mexicans and could have been more dangerous if the rebels
> could have gained more assistance from the U.S. or elsewhere.
> (Correspondence, September 1979)

By the middle of April the revolt was six weeks old. To most observers it was becoming increasingly apparent that the movement had
about run its course. And while the rebel forces in the Sonora sector
remained organized under a general officer, such was not the case
along the Texas border. The growing number of fragmented bands of
rebel soldiers appearing in this region became a matter of great concern for the United States military strategists. A spokesman for the
Eighth Corps Area stated on April 14 that the approaching cessation

of organized hostilities "is complicating matters and making more nec-essary than ever careful protection of the border. . . . Leaderless sol-diers are very apt to turn brigands and, in desperation, attempt ma-rauding attacks on American border towns and ranches." Alluding to the general fear syndrome that still gripped those living in isolated settlements along the Rio Grande, he explained further that "Villa's famous raid on Columbus, New Mexico, is still fresh in the minds of officers responsible for border protection" (*San Antonio Express*, April 14, 1929).

Dwindling rebel prestige throughout the Mexican military estab-lishment meant gains for the federal forces. Almost simultaneous with the arrival of the Ninetieth Attack Squadron at Fort Huachuca, a rebellion erupted for the second time within a month at Ojinaga, Chi-huahua, across the Rio Grande from Presidio, Texas. On Sunday night, April 8, the Ojinaga garrison revolted, assassinated the two rebel commanders, and proclaimed its loyalty to the federal govern-ment of Mexico. Since Ojinaga was in an area of strong rebel sympathy, it was feared that this 150-man federal force, although "well equipped with rifles, machine guns, and ammunition," would again be chal-lenged. Late Monday the garrison braced for a "possible attack by the small bands of rebels, known to be operating in the vicinity of San Carlos, near the village of Polvo [now Redford, Texas], about 49 miles south and east of Ojinaga" (*San Antonio Express*, April 9, 1929).

The anticipated attack on Ojinaga never occurred. Late the follow-ing Thursday afternoon two troop trains arrived from Chihuahua City, bringing reinforcements to secure the garrison. The commander an-nounced that, with his augmented force of approximately eight hun-dred soldiers, the threat of a rebel attack had ended. Although "three bands of rebel horsemen were reported in the vicinity of Ojinaga dur-ing the last week," that town expected no further trouble as "the revo-lutionaries were breaking up in small bands and disappearing" (ibid., April 13, 1929). Two days later one of these small bands of renegade soldiers appeared beneath the great cottonwood tree across the Rio Grande from Elmo Johnson's trading post.

The failure of the rebels to gain control of the northern ports of entry hastened the end of the conflict. Without foreign imports, Escobar could not supply his army. This, however, did not apply to his air force. Of Escobar's entire war machine, only the Yankee Doodle Escadrille maintained a constant source of foreign replacements. As rebel losses mounted, their agents in the United States continued to procure additional aircraft. Eager suppliers would land the planes at either the Phoenix or Tucson airports during the night, where rebel aviators awaited to ferry them to some remote airfield across the border. This practice continued until almost the end of the conflict.

When the Department of Justice agents learned of this clandestine activity, they began keeping the rebel ferry pilots under constant surveillance. Finally, as they were preparing to transport some replacement aircraft to Sonora, the federal agents quietly moved in. As a result, most of Escobar's veteran pilots involuntarily abandoned the rebel cause for the duration. The agents first arrested Pete Stanley at the Tucson airport. In quick succession Jack O'Brien, Pat Murphy, Phil Mohun, Richard H. Polk, and W. L. Fields, a recent recruit with Central American revolutionary credentials, were taken into custody at either Phoenix or Nogales. Journalist Tom Mahoney reported that "all were charged with violation of neutrality and the presidential arms embargo and lodged in Tucson, Ariz., jail." When a United States district attorney finally admitted his inability to prove where Stanley was going with the plane, "all were released except Mohun, who was transferred to El Paso and freed a few days later" (Mahoney, 1929: 75). Following the breakup of the Yankee Doodle Escadrille, its members quickly dispersed to pursue a variety of noncombatant occupations.[6]

As April drew to a close, conditions in Sonora appeared to be following the pattern developing along the Texas border. With the rebels' hope of receiving outside aid virtually abandoned, desertions became widespread. While small groups of rebel soldiers surrendered daily to federal forces, their former commanders began seeking political asylum in the United States. Gens. Francisco Monzo and Benito Bernal led the exodus. They crossed into the United States on April 12 and

6. On March 3, 1930, a grand jury in Tucson, Arizona, filed a seven-count indictment against Gen. José Gonzalo Escobar, several of his generals, and former members of the Yankee Doodle Escadrille. They were charged with unlawfully exporting arms and munitions of war from the United States to Mexico. The case was dismissed on May 25, 1932 (Los Angeles Federal Archives and Records Center, Regional Archives Branch, Records of the Twenty-first District Courts, Arizona, Tucson Division, Criminal Case Files C-4452).

One of the group's members, Phil Mohun, made his way to the West Coast where he reportedly found employment as a stunt pilot during the filming of Howard Hughes' World War I aerial saga, *Hell's Angels*. Polk eventually returned to his home in Nashville, Tennessee, where he practiced law from 1939 until he retired in 1976. Gustavo Salinas joined the Mexican pistol team and participated in the 1932 Olympics in Los Angeles. He explained his return to Tom Mahoney: "I just told them [the Mexican federal authorities] I was sorry and they let me come back" (correspondence, Mahoney to Polk, May 12, 1977; copy in author's possession).

were followed on April 20 by Gen. Francisco Urbalejo.

By April 27, the rebel forces had arrived at the final week of their brief crusade. Other than the small pockets of rebel soldiers positioned along the Arizona border, the last remnants of an organized army, which less than two months before controlled nearly half of Mexico, were reported separated and dispersed in southern Sonora (*San Antonio Express*, April 28, 1929). In the northern sector, General Almazán led his federal army through Pulpito Pass on the afternoon of April 28 and deployed his right flank to block the retreat of the rebels abandoning Agua Prieta.

As the federals drove toward a victory that now appeared imminent, battlefront reports indicated the federal air force formed the cutting edge of the operation that sent other prominent rebel leaders scurrying for the border. "Following the rout of the rebels by the federal aeroplanes in Southern Sonora," one official observer stated, "many of the leaders, including the Topetes and General [Francisco] Bórquez, went to Nogales, from whence they crossed into the United States." Then noting the landmark role of combat aviation in the rebellion, he continued, "It was probably the first time that [effective] aeroplane bombardment was made use of in a Mexican revolution, and during the last days of the rebellion it apparently had a pronounced effect" (*Foreign Relations*, 1943:3:424–425).

Witnessing the gradual disintegration, Immigration Service Inspector R. M. Cousar reported that

the crumbling process continued. At 9 P.M. on April 29, 1929, General Fausto Topete, the Commander in Chief of the Rebel Forces in the west coast states of Mexico, accompanied by Colonel Ricardo Topete and General Garcia, crossed into the United States and applied for admission at the immigration office. It was obvious that so far as Nogales, Mexico was concerned, the end of the Revolution was near. (May 7, 1929, RG 85)

Fighting, however, continued on the following day, April 30, gaining in intensity in the Nogales sector. At about ten o'clock in the morning a formation of three federal airplanes from the Naco base circled Nogales, Sonora, and dropped twelve bombs on the border village. There were no casualties, but the explosions and the return fire of the rebel guns created panic throughout the civilian section of Nogales. Cousar stated further that

the rebel forces opened fire upon the planes and stationed a large number of heavily armed soldiers—many of them with machine

Ground crews service Mexican federal airplanes at the Naco base for a mission against rebel positions along the Arizona border. The three airplanes at the left are United States civilian types, two Stearmans and a Travel Air; plane at right appears to be one of the Vought "Corsairs" purchased by the Mexican government prior to the outbreak of the Escobar Rebellion. Courtesy National Archives.

guns—along International Street, immediately adjacent to the Boundary Line. Great excitement prevailed on both sides of the line. Pandemonium reigned in Nogales, Mexico and the residents thereof, panic stricken and fearing that further aerial attacks would follow, rushed by thousands through the regular crossings into Nogales, Arizona for refuge. (Ibid.)

The exodus continued until three o'clock that afternoon, when it was estimated that ten thousand refugees had crossed to the American side for safety. The immigration officers on duty at the Arizona port of entry, under orders from the inspector in charge, "did nothing to prevent them from coming to Nogales, Arizona . . . and no one, regardless of race or nationality . . . was denied refuge" (ibid.).

Fear of additional air strikes from the federal air force plus the realization that the rebel cause was lost prompted the remaining rebel leaders to consider an alternative: surrender. At noon on April 30 a delegation headed by General Cocheu met with a group of rebel officers on the boundary line to negotiate the terms of surrender.[7] The rebel officers offered to surrender Nogales, Sonora, to the federal government "provided the Mexican Consul would, in the presence of General Cocheu, promise in the name of the Mexican Government that none of the Rebels would be executed or molested and would be furnished food and back pay" (ibid.).

Explaining that such a request required executive decision, Mexican consul Aveleyda agreed to discuss the matter by telephone with President Portés Gil in Mexico City and return later that day with his reply. At three o'clock that afternoon, all parties returned to the international line and resumed the conference. Aveleyda reported he had talked with the president, who authorized him to promise in the name of the Mexican government that

> the Rebels who then surrendered would not be executed or molested; that they would immediately be furnished with food; that they would receive some back pay . . . ; that the rebel privates and noncommissioned officers would be received into the Mexican Federal Army if they wished . . . ; that none of the Rebel officers would be taken into the Army but that the lives of those who then surrendered would be spared; and that they would not be molested. (Ibid.)

After accepting verbally the surrender terms, Aveleyda, Cocheu, and other members of the negotiations party crossed the international line to Nogales, Sonora. There the rebels surrendered the city to Aveleyda as the official representative of the Mexican government. Later that afternoon Gen. Lucas González arrived with his staff from Naco to assume command of the federal forces at Nogales. With the last remnants of the rebel forces under federal control, the revolt was over.

United States military units assigned to the Arizona sector during the emergency received orders to return to their permanent posts immediately following the cessation of hostilities. On May 7, 1929, Maj. Gen. William Lassiter, commanding general of the Eighth Corps Area, stated that "the Mexican border will revert to its normal mili-

---

7. Other members of General Cocheu's delegation were a Colonel Shipp, Col. Alexander MacNab (Ambassador Dwight Morrow's military attaché), American consul Altaffer, Collector of Customs Edwards, Mexican consul Aveleyda, and Immigration Inspector in Charge R. M. Cousar.

tary status as soon as the troop movements are completed" (*San Antonio Express*, May 8, 1929). The Ninetieth Attack Squadron, Third Attack Group, received orders for "staggered" return flights to the Fort Crockett airbase. "We flew back in flights of twos and threes over a period of several days," Colonel Rose recalled. "We didn't want the rebels to think we had all pulled out, just in case they might decide to start something again" (interview, January 28, 1980).[8]

General Escobar was not present at the surrender negotiations. After proclaiming himself provisional president of Mexico prior to leaving Juárez on April 9, his whereabouts had become a subject of wide conjecture, as he was reported at several different places at the same time. Prior to the surrender, his relatives in the United States were reported making plans for the insurgent general's escape to that country. California rancher John C. Uhrlaub confirmed that report on May 6, stating that Escobar and Gov. Fausto Topete had flown to a ranch near Gila Bend, Arizona, where they conferred with Governor Rodríguez of Baja California. When the conference ended Escobar requested permission to recross the border into Baja California, but Rodríguez refused. Following that rejection, Uhrlaub reported, "Escobar became so angry I thought there was going to be trouble" (*San Antonio Express*, May 7, 1929).

In discussing the rebels' defeat with Uhrlaub, Escobar "blamed lack of munitions and supplies for the downfall of the revolution. 'We had plenty of money . . . but the United States was against us. We tried repeatedly to get munitions across the border but failed because of the strict guard maintained by the American border controls.'" Defeat, however, appeared not to deplete the general's finances. "Escobar appeared to have plenty of money," Uhrlaub explained, "when he arrived at Gila Bend" (ibid.).

Escobar's movements after leaving Gila Bend remain as much a mystery as do his prior travels. He eventually made his way to Canada, where the profits of rebellion enabled him to live quietly, not to say retiringly, in a manner befitting a gentleman of means. This became a source of great annoyance to ex-president Calles. He maintained that Escobar "took a little matter of a half million dollars with him when he beat a firing squad to the border" ("How General Escobar Fought,"

---

8. Other units assigned to the Arizona sector during the Escobar Rebellion included the Eighty-second Field Artillery (horse) and the Second Cavalry, based at Fort Bliss, Texas, the Tenth Cavalry, from Fort Huachuca, Arizona, and the Twentieth Cavalry, based at Fort D. A. Russell, Wyoming (*San Antonio Express*, May 8, 1929). (Camp Marfa, located near Marfa, Texas, was subsequently redesignated Fort D. A. Russell.)

1931: 2). Mrs. Concepción Escobar apparently disagreed with Calles'
modest appraisal of the general's "cannon ball." In January 1931, she
filed an annulment suit (later dropped) for "half of the General's estate, and seizure before judgment has been placed on $1,000,000 de-
posited in banks here [Montreal] in his name." (*New York Times*, January 14, 1931).[9]

Canada remained Escobar's home away from home for more than a
decade. All attempts to extradite him failed and he rejected as well
President Lázaro Cárdenas' general amnesty offered the ex-rebels in
1936. He continued to enjoy the rewards of his prepaid exile until he
returned voluntarily to his homeland. On Sunday, May 24, 1942, the
former rebel general re-entered Mexico at El Paso "to offer his services
to the country in time of national emergency" (*El Paso Herald-Post*,
May 26, 1942).

---

9. See also "Patriots in Exile," *Esquire* (October 1937). Former *El Paso Post*
city editor Tom Mahoney interviewed General Escobar during his Canadian
exile. This interview appears in the *Buffalo Times*, November 1, 1934 issue.

Entr'acte

The two-month military debacle known as the Escobar Rebellion remains a blot on Mexico's frequently tragic yet colorful history. The loss in both human and economic resources contributed little to that country's three-decade struggle to institutionalize the political process. Power, prestige, and economic self-interest remained the motivating forces in a military campaign that claimed more than two thousand lives, repeatedly violated United States territory, and reignited a latent fear syndrome that would significantly affect the relationship of the two peoples that meet at the Rio Grande.

When examining Mexico's military history in broad perspective, the Escobar Rebellion casts an aura of subtle optimism; compared to other Mexican civil conflicts it appears as a watershed encounter. Casualties of the fifty-thousand-dollar cannon balls symbolize that country's revolutionary past, while the reforms inaugurated by Calles and continued during the Portes Gil and Ortiz Rubio governments forecast a new beginning. The more than half-century of domestic tranquility that has prevailed since the end of the Escobar Rebellion documents this change. (Dissident elements, however, continued to threaten Mexico's political stability. Civilians did not gain control of the government until after the bloody elections of 1940, when the revolutionary generals were finally stripped of political power.)

Close proximity to an event frequently distorts a viewer's perspective. That the Escobar Rebellion represented Mexico's last major spasm of violence understandably escaped recognition in 1929. By then it was a valid and well-documented assumption that guerrilla warfare would follow in the wake of revolution. The past contained a warning for the present. In 1915, under similar conditions, the Villistas had turned to banditry, looting, and extortion in the northern states, where policing was poor and governmental control nominal. Irked by United States recognition and support of Carranza, twenty-five Villista generals declared war on the United States in February 1916. This, plus the January execution of sixteen United States engineers and the March 1916 assault on Columbus, New Mexico, resulted in a United States punitive expedition, led by General John J. Pershing, against Villa. (Lieuwen, 1968: 34)

To those who occupied that remote wilderness that stretched along the international boundary, history had taught its lessons well.

The reign of fear touched off by Pancho Villa's rebels during the previous decade and rekindled more recently during the Escobar uprising was destined to continue. Thus the band of armed horsemen that Elmo Johnson watched across the Rio Grande that April afternoon in 1929 added confirmation—erroneous or otherwise—to the continuity of history. The implication was unmistakable; another phase in the usually peaceful, sometimes volatile, but always delicate relations between the two peoples who meet at the Rio Grande was about to begin.

"Most everyone in the lower Big Bend was aware of the airfield and knew it was there for protection."   —Hallie Stillwell

Scene 2.

In Search of Border Security: The Airfield at Johnson's Ranch

"... he fired at the bandits"

CHAPTER 5

Elmo and Ada Johnson's Big Bend ranch and trading post, which they established in 1927, occupied one of the most remote areas in the American Southwest. Located approximately 150 miles south of Alpine, Texas, the nearest urban center, they lived in almost total isolation. A few widely scattered Mexican families lived across the Rio Grande in Mexico, and Wayne Cartledge, their nearest Anglo neighbor, operated a farm and trading post sixteen miles upriver at Castolon. To the Johnsons, the appearance of a band of Mexican horsemen on the banks of the Rio Grande that April afternoon in 1929 was just cause for alarm.

Details of what occurred during the next few hours have been blurred by the passage of time. All participants are deceased. But one fact remains abundantly clear: Elmo and Ada Johnson came face to face with the Escobar Rebellion.

Subsequent investigation revealed that the band of thirty to forty mounted men were some of the remnants of the rebel cavalry that had escaped the federal slaughter at Jiménez. The battle ended on April 3, 1929, and during the following week the horsemen made their way slowly toward the Rio Grande. Being rebel deserters as well as fugitives from the federal army, they chose the most likely path of escape— toward the north. They traveled approximately 160 miles across the sparsely populated Chihuahuan desert; when they reached the Rio Grande and made camp in a dense mesquite thicket adjacent to the Johnson ranch, they were in one of the most isolated regions in northern Mexico. The nearest Mexican officials in the Big Bend area were four customs and immigration officials stationed at Santa Elena, a small village across the Rio Grande from Castolon. Since these officials seldom patroled the area near the Johnson ranch, the rebels could not have selected a safer hideout (Smithers Collection).

After establishing camp, the renegade soldiers turned to their immediate needs: medicine and food. Some of the men had been wounded at Jiménez. One sustained an abdominal wound and developed gangrene during the long journey on horseback. Two *curanderos* who lived about two miles downriver from the campsite were called to tend the wounded men, who subsequently recovered (ibid.) (*Curanderos* are native doctors without formal training who administer medications made from native sources such as herbs, plants, and insects.) Their next most pressing need was food, and for this they turned their attention to the isolated structure across the Rio Grande: Johnson's trading post. Although they may have embarked on this mission with peaceful intent, the episode ended with gunfire.

Sometime during the afternoon of April 11, 1929, two or three of the rebel soldiers forded the Rio Grande and came to the trading post asking for food. Johnson obliged, giving them some beans and corn. He told them when his herder returned later that afternoon he would also give them a goat. It remains unclear whether they returned that afternoon or not. Johnson would, however, meet the Jiménez survivors again (ibid.; Elmo Johnson, interview, June 3, 1966).

Elmo and Ada Johnson maintained their riverside vigil until darkness obscured their vision of the rebel campsite. Activity continued across the river; they could see silhouettes of human forms moving around a campfire. The Johnsons finally abandoned the watch, ate a late supper, and retired much later than usual. Around eleven o'clock Johnson was awakened by the sound of horses' hoofs followed by a knock on the door. Bandits, he thought, did not knock on doors. He went to the door, opened it partway, and saw the dim outline of two human forms standing nearby in the darkness. They spoke in Spanish: they wanted to buy some tobacco. Johnson was understandably suspicious; he seldom transacted business at night at the trading post. Nevertheless, he told them to go to the other end of the building where the commissary was located. Then, making his way across the large covered patio, he entered the commissary from a back door and lighted a kerosene oil lamp. Johnson then unlocked the door, admitting his two nocturnal customers.

They made their purchases, tobacco and cigarette papers, which they paid for with Mexican coin. As they proceeded to roll their cigarettes, Johnson viewed his customers in the dim lamplight. Both were Mexican; their appearance reflected the strain of battle and the torturous 160-mile horseback ride to the riverside campsite. Both were armed. Johnson had not bothered to pick up his pistol when he left the bedroom. This was not necessary; he kept a loaded rifle in the commissary.

As the two rebel soldiers smoked their cigarettes in silence, Johnson noticed they were growing nervous and uncomfortable in his presence. This further aroused his suspicions as to the real purpose of their late night purchase. At this point he must have begun to suspect that his worst fears were about to be realized. These two men bore him no favor. As Johnson turned slowly and reached for his rifle behind the counter, the two rebels fled from the commissary into the darkness. According to W. D. Smithers' newspaper account of the event,

> Johnson followed them to the door and heard two shots fired near by. About the same time he discovered that some other men were driving his cattle and goats . . . toward the river. . . . Standing on the bank above them, he fired at the bandits, who took off hurriedly, leaving the animals at the edge of the river. . . .
> Mrs. Johnson, who is also an expert marksman with the rifle, stays on at the ranch in spite of the insistence of friends that she is in constant danger. (*San Antonio Light*, April 30, 1929).[1]

Photographer-journalist W. D. Smithers, who represented the *San Antonio Light* and Underwood and Underwood News Service, had been making monthly trips through the Big Bend since 1922, gathering material for news stories, primarily on military activities along the Mexican border. By coincidence, he arrived at Castolon the morning following the raid on the trading post. On learning of the incident, he went out to the ranch, interviewed the Johnsons, made some photographs, and returned to San Antonio the following day to file his story. Texans would soon be reading that the era of banditry had returned to the Big Bend country.

By the time Smithers reached the trading post, Johnson had already reported the matter to border patrolmen George Dennis and Shelly Barnes, who drove to Castolon to report the raid to Chief Patrol Inspector Earl Falis at Marfa. Falis, in turn, forwarded the information to the post commander of the First Cavalry Regiment at Camp Marfa.

---

1. Smithers later questioned some aspects of Johnson's account of the event based on an investigation the Border Patrol reportedly conducted into the matter. The methodology the Border Patrol employed, however, remains questionable. Sending a local Mexican into Mexico to interview renegade rebel soldiers about their involvement in an international incident obviously lacked validity. A search failed to locate the report of that investigation (correspondence in my files).

Later that morning 1st Lt. Hugh F. T. Hoffman received Special Order No. 56 instructing him to take "F" Troop to Castolon, establish a base camp there, and launch an extended patrol of the lower Big Bend area (RG 394).

Because of the nature of the emergency and the time required for a troop of cavalry to march the 116 miles to Castolon, 2d Lt. Harry W. Johnson received verbal orders "to take my platoon of 'F' Troop, plus one section of machine guns and to move to Castolon by truck, to protect [the] American trading posts from Mexicans" (correspondence, March 2, 1980). Johnson recalled they proceeded via Alpine and Terlingua to Castolon, arriving there late in the afternoon. The young lieutenant established temporary guard posts to protect the Cartledge property and awaited the arrival of the livestock to proceed with the patrol assignment. Meanwhile, the remainder of "F" Troop plus the pack train was marching overland toward Castolon. This unit, consisting of four officers, eighty-nine enlisted men, and ten specialists, left Camp Marfa on April 12 and arrived at Castolon at two o'clock on the afternoon of April 15, covering the 116 miles in approximately forty-eight hours (strength returns, RG 407).[2]

Upon the arrival of the remainder of "F" Troop at Castolon, Lieutenant Hoffman divided the force, dispatching river patrols in two directions. He instructed Lieutenant Johnson to investigate conditions at Johnson's trading post (the two men were not related), and continue on downriver searching for the bands of renegade soldiers reported recently in that area. Hoffman proceeded upriver with his unit to Lajitas. The following day, Lieutenant Johnson

> left a small detachment and the machine gun section at Castolon, and with the rest of my platoon, took off for Johnson's ranch and trading post. (I had the pack train with me, as I recall, 52 mules.) Things were quiet at Johnson's, although there had been some Mexicans lurking across the river. I left several men at Johnson's, and with the rest of my platoon and the pack train, took off for Glenn Springs and old Boquillas. It was a terribly hot day and two of our pack mules dropped dead just out of Glenn Springs. (Correspondence, March 2, 1980)

---

2. The cavalry marched at widely varying speeds depending on terrain and duration. Three to six miles per hour was considered average speed for long-distance marches in the Big Bend region. The specialists listed are radio specialists, medics, and members of the veterinary detachment.

Members of "F" Troop, First Cavalry Regiment, relax on patio with Elmo and Ada Johnson following the April 1929 bandit raid on the trading post. Everyone is armed. The Johnsons each hold rifles, cavalrymen's Springfield rifles are stacked, and the Browning automatic rifle is assembled on the floor ready for action. Cavalrymen are unidentified. Courtesy Smithers Collection, Harry Ransom Humanities Research Center, University of Texas at Austin. Hereafter cited as Smithers Collection.

At Glenn Springs, Lieutenant Johnson met two Texas Rangers, Bob Pool and Ray Miller, who were pulling a two-horse trailer behind a car. They agreed to take the packs from the two dead mules and deliver them later at the cavalry post near Marfa. Difficulties facing the young cavalry lieutenant, however, continued to multiply. In addition to the heat, providing forage for the animals in that arid climate became a critical factor in maintaining the patrol schedule. The local Mexicans provided a solution to Johnson's problem. A few days later they began turning up with burros loaded with hay. The men also encountered dietary problems while on patrol, stemming largely from a steady regimen of "hard tack and corned beef packed in 1917." Johnson, however, resolved this matter personally, explaining, "Luckily, I took a 12 gauge shotgun with me and there were cottontail rabbits galore" (ibid.).

The patrol continued on to Boquillas, carefully surveying the south side of the Rio Grande "to spot any groups of Mexicans and parallel their movements." Finding conditions stable there, the unit remained at Boquillas one day and

> started up river toward Johnson's Ranch, now moving at night due to the terribly hot days. We moved through Hot Springs and on a trail along Mariscal Canyon. The load on one of the pack mules bumped an over-hanging rock and the mule and the load went tumbling about 50 feet into the river. We were traveling single file. We stopped and with lariats we "snaked" the mule and load back on the trail and we went on our way. (Ibid.)

Johnson's platoon returned to Castolon and, after a short rest, continued to scout the border periodically, mainly in the area of Johnson's ranch. Finding no further evidence of border incursions, the entire troop received orders on April 25 to return to Camp Marfa.

While "F" Troop patrolled the Rio Grande between Lajitas and Boquillas, the backwash of revolution spilled over into other regions of the Big Bend country. Following the federal takeover of the Ojinaga garrison on April 8, 1929, United States immigration officers arrested fifteen Mexican rebels on the Texas side of the river in that region. Those arrested included Ramón Armendárez, former mayor of Ojinaga, and Jesús Gallegos, a former justice of the peace.

Although the patrols conducted by "F" Troop reported no sightings of renegade soldiers near Johnson's ranch, Lieutenant Johnson acknowledged the potential danger still faced by those living along the Rio Grande. "As for the nature of the emergency," he wrote, "any isolated stations, such as Johnson and Cartledge had, were bound to be

sitting ducks for any marauding groups of Mexicans." Johnson also recognized the social duality of the area's Mexican population and its impact on the adjacent Anglo settlements. Crediting the predominantly
stable segment for countering the violence of those prone to rob and
plunder, he added, "The local Mexicans depended a great deal upon
these trading posts and farms for their livelihood, and they were loyal
to them and tended to protect the posts from marauding groups" (ibid.).

Although Elmo Johnson would experience the security provided by
his loyal Mexican neighbors, he continued to live with the fear of sinister forces lurking just across the river. The United States military
shared this fear. Even before the cessation of hostilities, plans were
already under way to establish an emergency air base in the Big Bend
area. This was the legacy of Mexico's revolutionary tradition; *violence
was certain to follow in Escobar's wake.* It would be W. D. Smithers who
would help switch the spotlight to Johnson's ranch, where the United
States Army Air Corps would ultimately locate an emergency landing
field. This unique facility would help alter the course of Big Bend history and bring an added measure of security to those living along the
international border. Of all the sites on United States soil touched
by the Escobar Rebellion, none sustained the lasting impact that
Johnson's ranch did.

After filing his story on the Johnson's ranch raid, Smithers went immediately to see his old friend, Col. Arthur G. Fisher, Eighth Corps
Area air officer, stationed at Fort Sam Houston.[3] As a photographer-
journalist covering Air Corps activities since 1920, Smithers had developed long-standing friendships with many Air Corps officers stationed at the San Antonio–area bases. Because of his pioneering work
in aerial photography, these officers frequently granted him official
permission to fly in Air Corps planes while on military-related assignment. From this experience, Smithers gained firsthand knowledge of
the Air Corps' emergency response capability.

In compiling his news story of the Johnson's ranch raid, Smithers
noted the time required for the horse cavalry to respond to the report
and recognized the need for a landing field in the lower Big Bend area.
Johnson's ranch offered a good site, and in the wake of this emergency
he believed the Air Corps administration would give careful consid-

---

3. The Eighth Corps Area air officer was the position formerly held by
Billy Mitchell. "In April, 1925, when his term as Assistant Chief of the Air
Service was up, he [Mitchell] was not reappointed. As a result, he was forced
to revert to his 'permanent' rank of colonel from his 'temporary' wartime rank
of brigadier general. He was transferred to San Antonio, Texas as Air Officer,
Eighth Corps at Fort Sam Houston" (Glines, 1963: 115).

eration to such a proposal. Smithers had previously mentioned this briefly to Johnson, who concurred with the idea. "Hell, I'll give 'em [the Air Corps] all the land they want," he assured Smithers. "I'll deed it to 'em, if necessary." There was one ulterior motive in Smithers' scheme. He would later write: "I [also] wanted one [an airfield] there to make many aerial pictures of the huge canyons and mountains" (August 27, 1975; Smithers Collection).

When Smithers related the account of the Johnson's ranch raid to Colonel Fisher and explained further the purpose of his visit, his host appeared surprised and asked, "How did you know that Washington wanted a landing strip in the Big Bend?" Smithers explained that he did not, but considering current border conditions in the light of the military's past problems in that area, he felt it was a wise decision. "I also stressed my beliefs," Smithers recalled, "that there may be other raids in the future as there were in the 1919 period." The colonel agreed, explaining that "he already had his orders from Washington to reestablish one of those old fields that the Border Air Patrol used in 1919 and 1920" (interview, August 27, 1975).

Fisher expressed an immediate interest in the Johnson's ranch site, which Smithers located on a large wall map in the colonel's office. Although Fisher lacked the authority to establish the landing field, he could order an inspection, evaluate the site, and channel his recommendations through the Eighth Corps Area command to the War Department in Washington. The officer who would make the actual inspection flight was Lt. Thad V. Foster, Eighth Corps Area assistant air officer, whom Fisher asked to join the conference. (Foster also carried the title of control officer.) When apprised of Smithers' proposal, Foster explained that he had already scouted the area unsuccessfully for a suitable location for a military landing field. Two days earlier he had flown to Glenn Springs trying to locate the landing strip formerly used by the 1920 Border Air Patrol. Ten years growth of greasewood and cactus, Smithers explained, had returned that site to its natural state (Smithers Collection).

The three men agreed that in terms of terrain and location Johnson's ranch was the preferred site for an emergency landing field where the Army Air Corps could land troops and base combat aircraft should another border emergency arise. Before making his final recommendation, however, Fisher decided to consult Maj. Robert J. Halpin, assistant chief of staff, Eighth Corps Area, and senior military intelligence officer. When told of the proposed Big Bend landing field, Major Halpin agreed to add his support to the project. He pointed out that in view of Mexico's volatile political climate, Johnson's ranch

Lt. Thad V. Foster (center), accompanied by Arch Miller, a Texas Ranger (left), and Ray Miller, San Vicente justice of the peace (right), carry flour sacks filled with dirt to use as ballast in the rear seat of the De Havilland DH-4. Foster, who flew W. D. Smithers to San Vicente to mark that site as an alternate Big Bend Air Corps facility, returned alone. Without added weight in the rear cockpit, the DH-4 flew "nose heavy." Note spare wheel lashed beneath the fuselage. Courtesy Smithers Collection.

would be an excellent location from which to gather information on activities along the Mexican border.

Before the meeting ended, Fisher instructed Foster to fly Smithers that afternoon to the Army Air Corps auxiliary landing field near Dryden, Texas. From there a Sergeant McCabe would drive him to Johnson's ranch in a truck, carrying gasoline, oil, a windsock, and everything needed to clear and mark the field and service the plane. Two days later Foster would fly to the ranch and inspect the site. (Fisher was obviously propelled by a sense of emergency. At that time Nogales and Agua Prieta, Sonora, remained rebel strongholds; surrender at Nogales was still one week away.)

Smithers and McCabe arrived at the ranch late the following day. Early the next morning they began removing all the large rocks and big clusters of petaya cactus from the landing area and inscribed a large letter **N** in lime on the north side of the field. Johnson, who had a road grader at the ranch, cleared a long strip of greasewood sufficient to accommodate Foster's De Havilland DH-4. Then they awaited the assistant air officer's arrival. Smithers recalled that about ten o'clock on the morning of April 24, 1929, they heard the drone of a Liberty engine and "ignited two smoke cannisters that made a lot of yellow smoke for Foster to see. He later said that he saw the large tin roof of Johnson's house before he saw the smoke." Smithers explained that "Lt. Foster was delighted with the location and everything about the field," and indicated there should be no problem in getting the landing site approved by the War Department (ibid.).

Smithers and McCabe left immediately in McCabe's truck for the Air Corps auxiliary field at Dryden, while Foster decided to remain overnight with the Johnsons. The following day he flew back to Dryden, picked up Smithers, and they returned to Fort Sam Houston (Dodd Field) to report their findings to Fisher. The colonel, in turn, forwarded his recommendation to the office of the chief, United States Army Air Corps in Washington, and awaited a response.

On that inauspicious occasion the Big Bend of Texas entered the modern air age. And while the size of the landing field would be enlarged and the condition improved over the coming months, the United States Army Air Corps had, in fact, established a beachhead on the northern banks of the Rio Grande. According to Smithers, this was demonstrated

two weeks later, [when] two Douglas transport planes [C1-Cs], each with 12 Infantry soldiers—each with full equipment—landed on the new field. They flew the 350 air miles . . . in less than 4 hours from Ft. Sam Houston. That flight was a warning to the followers of the Escobar Revolution that there would be no reoccurrence of Mexican border trouble in 1929 as there was in 1916, because of Mexico's [previous] revolutions. . . . By planes, soldiers could be there in the same number of hours that it would take the horse cavalry [days] to get there. (Ibid.)

While Smithers' interpretation of the flight was personal, it reflected the official military view of conditions along the Mexican border and the most expeditious method by which they could be resolved. A show of military force would be a warning to rebels as well as a deterrent, and this time the United States Army Air Corps would assume a major share of the responsibility.

The preliminary use of the Big Bend airfield, still pending official authorization, demonstrated the high priority the United States Army Air Corps accorded the border emergency. The successful deployment of combat-ready troops to the remote airfield no doubt further insured approval of Colonel Fisher's recommendation. About two months later, on July 6, 1929, Lieutenant Foster flew back to Johnson's ranch carrying the official authorization for the landing field and a one-dollar-a-year lease agreement for the Johnsons to sign. Only six people were present for the field's dedication. Sergeant McCabe drove down from Dryden in his truck, and Smithers arrived by car shortly before Foster landed in a De Havilland DH-4, No. 167, accompanied by a Private Ashworth. The six people retired to the covered patio of the Johnson's ranch house for the lease signing (RG 18).[4] Following the brief ceremony, Foster inscribed the first entry in the field register: "7/6/29 - DH 167 - Lt. T. V. Foster - Pvt. Ashworth - 8CA."[5]

The Johnsons kept the register on the patio for the next fourteen years and it became a unique document in the annals of military aviation history. Many young lieutenants who inscribed their name, rank, type of aircraft, and home base in the field register subsequently elevated that rank in the United States Army Air Corps (and later the Air Force) to command-level assignments during and after World War II. Two of the more notable inscriptions are those of Gen. Nathan F. Twining, later chairman of the Joint Chiefs of Staff, and United States Army Gen. Jonathan M. Wainwright, hero of Bataan, then post commander at Fort Clark, Texas. Entries of the various types of aircraft flown to Johnson's ranch indicate the Big Bend airfield also became an unlikely showcase for military aircraft development. Slow observation type biplanes, constructed largely of fabric-covered wooden wings and steel tubing fuselages, are gradually replaced by larger, faster, and more combat efficient all-metal fighter, attack, and bomber types. The final

---

4. United States Army Air Corps lease agreements, correspondence relating to subsequent Works Progress Administration fund applications, and G-2 Intelligence reports, U.S. Army Continental Commands, constitute the only official documentation found on the airfield at Johnson's ranch. Other Air Corps (and later air force) records contain no references to this facility. One retired air force officer explained, "It was conducted as a sort of moonlighting operation. You won't find many official records of that operation" (Deerwester, interview, February 11, 1971).

5. DH 167 identified Foster's De Havilland DH-4 aircraft numbered 167; 8CA is the abbreviation for the Eighth Corps Area. Sul Ross State College president Bryan Wildenthal and college librarian Dudley Dobie later acquired the airfield register from Elmo Johnson. It is now in the Archives of the Big Bend, Sul Ross State University, Alpine, Texas.

pages of this yellowing document reflect accelerating developments in the aircraft industry, as the Army Air Corps began building an air armada commensurate with changing world conditions. (All further reference to air traffic at Johnson's Ranch are taken from the field register.) [6]

Although the airfield located on the Johnson ranch represented the Air Corps' first permanent installation in the lower Big Bend, those living along the Rio Grande were well accustomed to the sights and sounds of military aviation. Following Pancho Villa's attack on Juárez in June 1919, Maj. Gen. Charles T. Menoher, chief of the United States Army Air Service in Washington, ordered eighteen De Havilland DH-4 airplanes to Fort Bliss to inaugurate an aerial border patrol. This original group, commanded by Maj. E. G. Tobin, patrolled the United States–Mexican border from San Diego, California, to Brownsville, Texas (Hinkle, 1967: 5). Aircraft based at Marfa patrolled the Rio Grande from El Paso to Sanderson. The eastbound patrol flew south to Presidio, turned east along the Rio Grande, and passed over the heart of the Big Bend country. In order to maintain closer liaison with the cavalry, the Air Service established a temporary landing field at the cavalry outpost at Glenn Springs. Calling that rough, rocky strip a landing field for aircraft, according to Smithers, was a gross exaggeration. "There never was, really, a field at all there. The 6th Cavalry, and later the 8th Cavalry, just cleared off a few rocks and some of the bigger petaya cactus, and chopped down some of the biggest greasewood." Smithers added, however, that the DH-4s were well suited to this type of military operation, as "those old World War I pilots could sit one down in just about any old place" (interview, August 27, 1975).

The DH-4s the Border Air Patrol used in the 1920s differed little from the one Thad Foster first flew to Johnson's Ranch nine years later. The design and quite possibly the actual airplanes were of World War I vintage, and their availability became a factor in retarding the growth of United States military air power well into the 1930s. This unique airplane represented the wartime mating of the British-made De Havilland airframe and the American-designed and manufactured Liberty engine. During World War I three American firms produced

---

6. United States War Department lease agreements refer to the site informally as the landing field at Johnson's Ranch. Works Progress Administration correspondence cites the facility as "Johnson Ranch Airport" and "Johnson's Farm Airport." The San Antonio–area Air Corps bases that cleared flights there referred to the landing field simply as Johnson's Ranch. The latter term, as applied to the military airfield, will be used throughout the remainder of this work.

4,846 DH-4s, and these became the first and only American-built planes to fly over enemy territory during the war (Wagner, 1960: 51). Following the war, this multipurpose fighter, bomber, and reconnaissance aircraft became the mainstay of the fledgling United States Army Air Service and remained in service for more than a decade.

By the late 1920s, the manufacturer had made some minor modifications in the DH-4 design (Boyne, 1979: 33–45). Steel tubing, for example, replaced the original spruce fuselage framework, and the fuel tank was moved forward of the pilot's seat, facilitating communication between the pilot and his gunner-observer, who sat in the rear seat. But visually the DH-4 remained unchanged. Its blunt nose, formed by the radiator that cooled the twelve-cylinder four-hundred-horsepower Liberty engine, gave the De Havilland its most identifiable characteristic, while the spare wheel tied beneath the fuselage suggested the afterthought of last-minute packing. It was, however, the twelve wooden struts secured by exposed guy wires that braced the fabric covered wings that created an unforgettable silhouette of fragile mobility.

The one DH-4 design characteristic that gave the Air Corps pilots major concern when landing on unfamiliar fields was the landing gear. The wheels, attached to an axle (called the spreader bar), allowed about a twelve-inch clearance above the ground. Any obstacle protruding that distance—grass, brush, rocks, snags—created a serious hazard during takeoff and landing. Lt. Elmer P. Rose, stationed at Brooks Field in 1929, experienced the nervous anxiety of landing a DH-4 at Johnson's Ranch before the field had been improved. Rose's principal concern was the native grass that covered the field and

> grew up, as a rule, about twelve to eighteen inches high. . . . Unfortunately, [the spreader bar] would catch in the grass, and unless you landed with your tail low, and used some power to keep your tail low, you might go up on your nose and break your propeller or damage the engine. So every landing was a sporting course. There weren't any two the same. (Interview, January 28, 1980)

As a navigational aid for the younger Air Corps pilots who first ventured into the Big Bend aerial frontier, Johnson placed two piles of rock about one hundred feet apart at the west end of the runway. "If you landed in between," Rose explained, "landed and touched down with your tailskid between those two piles of rock, and kept it down with some propeller blast from your engine, then you were all right" (ibid.).

The DH-4 emerged as both the visual symbol and performance model of post–World War I military aviation. This was an air age when slow speeds, short cruising ranges, and flights of unpredictable outcome characterized aircraft performance. Of all types of aircraft then in service, none was more typical than the DH-4. Speed and stability were not its greatest assets. "We used to say it cruised, took off, and landed at ninety," explained Gen. William L. Kennedy, who made his first DH-4 flight to Johnson's Ranch in 1931. According to Kennedy, the plane actually cruised around ninety miles per hour and landed under normal conditions at around fifty miles per hour. "You brought it in faster than it had to be because of its qualities [flight characteristics]. . . . it wasn't too stable." Kennedy added that without this additional landing speed the DH-4's narrow undercarriage and longitudinal instability made ground looping a distinct likelihood (interview, September 11, 1979).

In addition to their marginal dependability, the Liberty engines created an additional problem. The roar of the engine was deafening for the pilot who occupied the open cockpit immediately behind the short exhaust stacks. Gen. Robert S. Macrum explained he put cotton in his ears before every flight because "the punishment to your ears was really terrific." Some of the cadets called it "sissy, sticking cotton in your ears every time you got in an airplane to fly. And they didn't. Well, you know how it turned out with them? Some of 'em are now deaf and got some partial disability when they retired." Then, pausing to recall his early Air Corps experiences at Johnson's Ranch, he added, "But I'm awfully glad I had the experience of flying in the DH-4s" (undated tape, 1979).

The first official inspection flight to the airfield occurred one month after the dedication on July 6, 1929. On August 5, Lieutenant Foster returned to the field in his DH-4, No. 167, accompanied again by Private Ashworth. As assistant air officer in the Southwestern Airways Section of the Eighth Corps Area command, it was Foster's responsibility to see that all facilities within the Airways Section were properly maintained and supplied. Johnson's Ranch was one of many landing fields for which Foster was responsible. And, ultimately, it became his favorite.

With official approval by the War Department as an Air Corps landing field, Johnson's Ranch became part of the Southwestern Airways Section. This was the organizational structure that administered the chain of military landing fields that linked the various bases within the five-state Eighth Corps Area command. During this period of short-range and frequently undependable aircraft, this network of closely spaced landing fields was essential for the long-distance operation of military airplanes.

The Corps Area constituted a service command, and the air officer in charge of the Airways Section administered the landing fields within the command area. Originally, a small complement of men was assigned to the military aviation section on detached service. Later, when the Airways Section became a designated unit, the men were assigned directly to that section for duty. The five-state area within the Airways Section (and the Eighth Corps Area)—Texas, Oklahoma, Colorado, New Mexico, and Arizona—extended roughly from Fort Crockett (Galveston, Texas) in the east to Yuma, Arizona in the west and from Fort McIntosh (Laredo, Texas) in the south to Hatbox Field (Muskogee, Oklahoma) in the north.

The landing fields within the Airways Section fell within three general categories, varying according to ownership, management, staff, and service accommodations. The largest number of these facilities were designated as emergency landing fields, meaning they were simply cleared and marked landing areas maintained for emergency use only. The Airways Section provided neither aircraft services nor personnel accommodations, and these fields were maintained either by local people under contract to the Air Corps or by military personnel stationed elsewhere within that immediate area.

Other airfields within the Airways Section offered a wider range of services for transient aircraft and crews. The auxiliary landing field near Dryden, Texas, for example, was owned and operated by the Air Corps (the land was leased) and was equipped with a hangar, overnight quarters for stranded airmen, a limited inventory of spare parts, and quarters for a noncommissioned officer, who, in some instances, qualified as an aircraft and engine mechanic. The Lordsburg, New Mexico, field, similar to the Dryden field, served mainly as a midway refueling stop between Biggs Field (El Paso), Texas, and Tucson, Arizona. Hatbox Field and the Tucson, Arizona, facility also provided a service that corresponded to the Dryden field, except they were operated in conjunction with municipally owned airports. The Midland, Texas, operation was similar, except a private commercial aviation enterprise owned that airport and shared hangar space with the Air Corps. Living accommodations were maintained for a noncommissioned officer who helped service and maintain transient military aircraft.

Hensley Field at Grand Prairie, Texas, represented a third type of landing facility. Operated in conjunction with the United States Army Air Corps Reserve, Hensley Field facilities included a hangar, fuel storage, classroom facilities, and a permanent allocation of military aircraft in which reservists received periodic training and maintained minimum proficiency levels as required by the Air Corps. Installations such as Hensley Field were administered by an officer (usually a captain), who supervised a minimum staff that maintained the aircraft and conducted weekend training sessions.

Such was the network of landing fields of which Johnson's Ranch became a part. Yet in terms of its location, facilities, administration, and purpose, it differed enormously from all other airfields in the Corps Area. Since its location placed it more than one hundred miles off an established airway, it could be considered neither an emergency nor auxiliary field in the strictest sense. The Army Air Corps established this riverside air base solely because of its strategic location, where combat troops could be landed and aircraft be deployed in case of a border emergency.

In addition, its facilities and administration were definitely unique. The landing area either equaled or exceeded that of most other stations in the Airways Section, while a ranch house and a trading post served as the field's operational center and provided overnight accommodations for the air crews. Yet the administration of the field became one of its most unusual aspects. This airfield came under the quaint civilian domination of Elmo Johnson, who sought neither rank nor authority but who ultimately filled a rare niche in the military hierarchy for which there was neither precedent nor counterpart. Unlike any other facility in the system, Johnson's Ranch evolved from entirely different needs and developed in a manner that carried its functions to the borderline of officialdom and sometimes beyond. Referring to the marginal nature of some of the field's activities, Lt. Col. Howard E. Rinehart, a former member of the Southwestern Airways Section, explained,

> There was just nothing else quite like Johnson's Ranch. . . . Basically . . . [it] was a loose operation. We had no manning authorization, no maintenance funds, no equipment allowances for it. . . . We pulled men from Dryden and Marfa on a temporary basis to operate the radio facility. Mr. Johnson, with a little "undercover" assistance from SW Airways, did the rest. (Interview, September 10, 1980; correspondence, November 10, 1980)

The facility, however, was very official, explained Col. John W. Egan, who first landed at Johnson's Ranch in 1931 on an emergency mission. Militarily "it was a recognized field for the purpose of border patrol . . . and the protection of citizens that lived along the Rio Grande" (interview, February 7, 1978). Hallie Stillwell, a long-time Big Bend resident, historian, and journalist, agreed: "Most everyone in the lower Big Bend was aware of the airfield and knew it was there for protection. . . . In fact, a bunch of us neighbors attended a fishfry there celebrating the event" (correspondence, November 6, 1978).

It was indeed an event to celebrate.

"... all we did was land in flat places"

CHAPTER 6

The establishment of the emergency landing field at Johnson's Ranch ended, at least temporarily, the threat of marauding bands of Mexican horsemen reportedly lurking along the Rio Grande. Neither Fort D. A. Russell nor the Big Bend airfield reported any border incidents subsequent to the Johnson Ranch raid. The presence of military aircraft operating out of Johnson's Ranch apparently ended any danger to Big Bend settlers following the Escobar Rebellion.

Of the seventeen landings recorded at the Big Bend airfield between July 6, 1929, and July 5, 1930, eleven were made by Lt. Thad V. Foster in his DH-4, No. 167, while on routine inspection flights. During this period, flight operations were conducted on a single narrow strip improved mainly by removing the brush and rocks. Following Foster's second flight to Johnson's Ranch on August 5, 1929, W. D. Smithers reported that the assistant officer "gave instructions as to the proper size of the area to be cleared and for its care. . . . The field, when completed, will be about three-quarters of a mile square" (*Alpine Avalanche*, August 16, 1929). Smithers recalled later that "they sent a man down there to enlarge the field to Army specifications . . . [and] left the grader there for Johnson to keep the field in shape" (interview, August 27, 1975).

When Col. John W. Egan first visited the Big Bend airfield in 1931, he thought "it was a pretty sorry job of clearing off the brush. . . . [Johnson had] hired some Mexicans to grub out the mesquite and the bushes that were on the ground. . . . It was pretty dangerous on tires" (interview, February 7, 1978). Although the condition of the field improved markedly over the years, its initial shortcomings were obvious even to civilians. This included attractive young ladies, for whose aerial companionship Foster displayed a pronounced predilection. Lois

Neville Kelley recalled a daring night flight from Alpine to Johnson's Ranch with Foster in the early 1930s:

> I remember how it looked when Captain Foster [he was a lieuten-
> ant then] flew me down the first time and how it slowly changed
> the next several years as the greasewood gave way to dragging.
> . . . And it was kept up rather well. No more torn wings and
> damages from rocks and bushes. . . . The added risk of danger in
> landing at night without lights at Johnson's, made this trip one of
> the most unforgettable of my life. [Johnson provided two lighted
> lanterns to identify the landing area.] (Correspondence, October
> 24, 1980)

Field size and surface conditions determined the type of aircraft the Air Corps dispatched to the Big Bend airfield; graded runways opened the way for more advanced equipment. Egan noted that he never tried to land a Douglas O-35 there until September 14, 1934. A twin-engine, high-wing, reconnaissance light bomber, the O-35 (also designated the B-7) carried a three-man crew and was considered a high performance aircraft in the 1930s. Powered with two six-hundred-horsepower engines, it operated at a maximum speed of 183 miles per hour and landed at approximately seventy-five miles per hour. Although the O-35's performance requirements may have taxed the new field's facilities, gradually improving conditions allowed other more advanced aircraft to be cleared to Johnson's Ranch. Johnson ulti-mately extended the runways to accommodate the Air Corps' fast-developing high-performance aircraft. By 1939 the field had three graded runways that measured north-south, 2,100 by 400 feet, north-west-southeast, 2,900 by 600 feet, and northeast-southwest, 4,200 by 900 feet. (For comparison, these specifications far exceeded those of the Austin, Texas, municipal airport in 1939, where two runways measuring only 2,000 and 3,000 feet served private and commercial operators while Fort Sam Houston's Dodd Field provided two crushed stone strips of 1,800 feet and 1,050 feet.)

Foster's primary concern was transforming the Big Bend airfield into a safe, all-weather operational facility to which emergency combat air-craft could be dispatched. Supervising the work at Johnson's Ranch as well as being responsible for the actual administration, maintenance, and supply of all emergency and auxiliary landing fields within the Corps Area was a demanding assignment that, according to his con-temporaries, Foster relished. Gen. William L. Kennedy, who knew Foster well, stated it was Foster's chosen "mission in life to fly that border and see to it that these stations existed, and that the flat places [sufficiently large for landing an airplane] were known. He was as-signed to the Corps Area, but flew out of Dodd Field in an old DH-4" (interview, January 19, 1978).

Colonel Egan remembered Foster as a very humorous fellow. "Just loved to fly. He spent his life in an airplane. Anytime of the day or night that he could fix up an excuse to fly, he had the airplane strapped to him and he was gone. He was a great pilot, very skillful." Foster was "not just foolin' around" with his airplane, Egan added. "Each flight had military purpose" (interview, February 7, 1978). This claim is fur-ther substantiated by the field register. Following Foster's second flight to Johnson's Ranch on August 5, 1929, he returned to the field on August 23, accompanied by a Lieutenant Commander Young, United States Naval Reserve Corps, and on October 1, again accompanied by Private Ashworth, Foster included Johnson's Ranch in an inspection tour of all emergency landing fields between San Antonio, Texas, and Tucson, Arizona. A flight of this magnitude made news in the late 1920s, and the *San Antonio Express* reported that "the route is dotted with emergency landing fields at intervals of 12 miles, each field being marked with concrete rings in the center. . . . Lieut. Foster is making the trip along the route by plane, and will land at each of the 50 fields" (*San Antonio Express*, October 4, 1929). Foster's familiar DH-4, No. 167, remained his vehicle for all administrative flight assignments throughout 1929.

Within the scope of this study Foster and his DH-4 airplane emerge as historical benchmarks, symbolic of a transitory phase in the devel-opment of the United States Army Air Corps. Both the man and his aircraft, each in a very specific manner, represent the interwar status of the nation's air arm. Foster, as an Air Corps pilot and officer, en-joyed both personal independence and public adoration not accorded fellow officers in other branches of the service. And he, like most people who flew airplanes in the 1930s, became a folk hero to a nation fascinated with men with wings.

Foster's airplane, the DH-4, a product of wartime emergency, repre-sented the newest branch of the national defense establishment, a corps unit held in constant check by a tradition-bound general staff unsure of the airplane's military potential but ever fearful of its chal-lenge. Combined, they—the man and his airplane—are essentially historical timepieces, products of a specific era and created by its unique conditions. Thus the man, the airplane, and the Big Bend air-field with which they became so closely identified emerge as symbolic links in the endless chain of cultural and historical change.

When examining the Johnson's Ranch experience in broad perspec-tive, Foster's influence stands paramount. It is particularly signifi-

Lt. Col. Thad V. Foster in 1942 when he commanded the Jackson (Mississippi) Army Air Base. Formerly the Eighth Corps Area control officer, Foster was the person most responsible for establishing and maintaining Johnson's Ranch airfield. Courtesy Lois Neville Kelley.

cant that his distinctive signature dominates the early pages of the Johnson's Ranch field register. His frequent entries may suggest more than the lieutenant's official responsibilities at the airfield.

Foster and Johnson very early became close personal friends. The friendship that developed between these two individuals influenced greatly the operation—official and otherwise—of the Big Bend landing field. According to Col. Howard E. Rinehart, who as a sergeant frequently flew to Johnson's Ranch with Foster, "the relationship between Elmo Johnson and the Corps Area—and between Johnson and the Air Corps—was a very loose, mostly verbal relationship that was maintained largely through the friendship between Thad Foster and Elmo Johnson." The two men formed "a close mutual admiration society. Both men had high regard for each other" (interview, September 10, 1980). In all probability, each recognized in the other personal qualities that they admired and individual characteristics that they attempted to emulate. Both were fiercely independent, highly individualistic, and their actions were based largely on personal choices.

While rebel is probably too strong a term to describe either Johnson or Foster, it does, however, suggest the character of the two men who administered the remote Air Corps landing field that overlooked the Rio Grande and Mexico beyond.

By whatever measure Foster's role in the Johnson's Ranch operation is evaluated, it must be remembered that while he was representative of the cadre of young flying officers that comprised the Army Air Corps of the 1930s, he was far from typical. Contemporaries categorize him as one of a kind. Claims that "Foster was a character" are also substantiated by his personal remarks and unofficial entries in the field register. Some of these tend to partially negate Egan's claim that Foster was "not just foolin' around." While he undoubtedly knew that Air Corps officers were accorded a wider latitude in personal conduct than his counterparts in other branches of the service, he must have also known that inviting young women to accompany him on inspection flights violated every applicable rule and regulation enacted by the United States Army Air Corps. Foster nevertheless made frequent trips to Johnson's Ranch with his girl friends. In addition, he exhibited the brazen audacity to enter their names in the register, frequently illuminated by his personal comments. A former aide, when asked if Foster was not risking court-martial by inviting women aboard Air Corps planes, exclaimed, "All the time! All the time! When they finished Thad, they busted the mold. He was different from anybody else." [1]

Foster lists his first companion as "Miss Clarissa Ryan," who was followed by "Mary King RCN [Red Cross Nurse?] & all that sort of ROT." Although the abbreviations following Mary King's name are not entirely clear, she is presumably the niece of Col. Joseph King, First Cavalry Regiment executive officer stationed at Fort D. A. Russell. Gen. Samuel L. Myers, a former member of that unit who knew both Foster and Mary King, stated, "I didn't know that he was hauling Mary King around in his aircraft, because it was very much against regulations then, as it is now" (undated interview). The aerial parade of Foster's female friends continued, however, with "Miss Moss," "Mrs. Holmes" of Dryden, Texas, and "Mrs. Lee King." The entry for "Miss Lois Neville" of "Alpeen" (Alpine, Texas) terminates Foster's itemized list of aerial companions. Something of a permanent nature apparently resulted from this relationship. In the contemporary vernacular, they became an item.

---

1. My source, who wishes to remain anonymous, recalled that on several occasions Foster entered his name on the flight clearance when a young lady was actually his passenger.

Some junior officers noted Foster's awareness of the authority vested in his assistant air officer/control officer title, and recall he relished the opportunity to exhibit his nonconformist attitude toward routine matters. Gen. Robert S. Macrum explained that shortly after arriving at Dodd Field in 1931 as a second lieutenant he was assigned operations officer of the day. During this tour of duty he saw Foster walk out to the flight line carrying his parachute. Assuming he was about to embark on a cross-country flight,

> I came over to him as he was getting into his plane, because I felt it was customary to send a flight plan to his destination. Well, he turned to me with a slight sneer in his voice, and said, "Don't you worry about me." So, I got no information about where he was going, or how long he would be in the air, when he intended to arrive, and at what time. So I didn't pay much attention to Thad Foster after that. But I'm not surprised at anything that he might do. (Undated interview, 1979)

The Army Air Corps in the early 1930s by its very nature permitted a high level of individual freedom, but according to Macrum, Foster pushed the libertine spirit to its limits.

> I envied Thad. He was just an independent operator. Just made up his own program on anything and everything on flying. And I suppose *when* he would go to work, and *if* he would go to work, because nobody kept a check on him. . . . His airplane was just a nice toy for him as he could fly anywhere at any time. . . . So he did a lot of flying around. Maybe he could have made some money, for all I know, by taking people up for rides. (Ibid.)

Although Macrum's reference to the commercial aspect of Foster's relaxed routine was a facetious exaggeration, his request that the control officer file a flight plan was a simple yet highly essential routine. During the early 1930s, after the operations officer approved a cross-country flight request, the pilot gathered the latest weather information from base operations, entered his name and the aircraft identification number in chalk on the operations board, and listed his destination and the date and time he planned to return. "And that's all there was to it," Gen. William L. Kennedy explained. "In certain instances you had to send a Western Union message back, but if there wasn't any place to send a message back from, well, obviously you didn't do it" (interview, January 19, 1978). As the Air Corps expanded, the flight clearance procedure became more sophisticated. Col. Elmer P. Rose recalled that "later on you had to get permission from the post operations officer in charge to land anywhere other than at your home base. You had to fill out a form showing the type of airplane, your name, the route you were going to fly, and whether or not you were going to return that day." But even with all of the proverbial military "red tape, they were still very lax about going out to Johnson's Ranch" (interview January 28, 1980).

During this period when the Air Corps operated marginally dependable equipment in a command area that lacked both air-to-ground radio communications and in-flight progress reports, some manner of aircraft control and accountability was mandatory. The flight plan at least provided base operations with essential information on all cross-country flight operations. These operations also occurred before the development of scientific weather forecasting and over terrain inadequately and frequently incorrectly recorded on the Air Corps' flight maps. This latter factor often resulted in navigational errors. Col. LeRoy Hudson, a former member of the Twelfth Observation Squadron based at Brooks Field, explained that the poor flight maps (they were railroad maps) resulted in "many incidents that were hilarious and some quite frustrating at the time" (correspondence, September 1979). Joel G. Pitts, Hudson's former colleague and retired Braniff International Airways captain, recalled vividly, and with some embarrassment, the source of the above statement.

During the Escobar Rebellion, Pitts was assigned transport duty, flying a Douglas C-1C between Dodd Field and Fort Huachuca. The Ninetieth Attack Squadron was temporarily assigned to that post, and Pitts supplied the unit with spare machine gun parts, tires, wheels, and other aircraft parts. On his first westbound flight he landed at the Dryden auxiliary field for fuel, and was ordered to remain there until a sandstorm in the El Paso area subsided. He received his clearance sometime in the afternoon, allowing him just time to reach Biggs Field before nightfall. As Pitts approached the El Paso area, he set his course in accordance with his map and began looking for the field, with which he was unfamiliar. When he arrived at the site indicated on his map, there was no field. With darkness approaching, Pitts circled the location and, finding no place to land, continued on to Fort Bliss and landed on the cavalry post parade ground. Although he realized his decision was not in accordance with his orders, he felt his initiative in seeking out an alternate site merited his superior's commendation. Unfortunately, this was not the case. "Major E. A. Lohman," Pitts recalled, "never let me forget that I landed at the wrong field" (interview, November 16, 1980).[2]

Col. John W. Egan, a frequent visitor to Johnson's Ranch in the early 1930s, reexamined his flight maps years later and noted the inaccuracies. "These maps don't even have Johnson's Ranch airfield on it," he explained. "They were printed years before Johnson's went in there. We were still using them in the thirties" (interview, February 7, 1978).

Recognizing the need for more accurate navigational data, the Air Corps inaugurated a program to correct the flight maps in use at the time Johnson's Ranch was established. In 1930, Lt. Joseph H. Hicks, commander of the photographic section at Dodd Field, and Lt. William L. Kennedy received orders from Washington to make on-site corrections of Air Corps navigation maps along the United States–Mexican border. At that time, according to Kennedy, the Air Corps was issuing the old Rand McNally railroad maps for pilot navigation, and they were totally unsatisfactory. Hicks and Kennedy embarked on their journey in a Fairchild YF-1 photographic aircraft and "landed at flat places all the way to California," Kennedy recalled. "It was really a process of map correction. We would land and talk to people. We put roads in and erased the things that were not supposed to be there." On the return trip they landed at Johnson's Ranch on July 7, 1931. The field was not graded then but was just a flat place, Kennedy remembered. "Back then, all we did was land in flat places. . . . That was the first time I met Elmo and saw the place. I was fascinated with it, of course" (interview, January 19, 1978).

Kennedy's fascination with the new airfield, the primitive appeal of the lower Big Bend country, and the indigenous charm of the Johnsons led to his scheduling many return flights to Johnson's Ranch. The uniqueness of this remote Airways Station also attracted many other young Air Corps pilots. During the next decade, hundreds of military personnel of all ranks sought out this riverside Airways Station for duty, training, fun, and recreation. Credit for this exodus from the San Antonio–area bases is due mainly to W. D. Smithers, who, following the dedication of the field, asked the Johnsons if they would welcome weekend Air Corps visitors at the ranch. Foster overheard Mrs. Johnson's positive response, and, according to Smithers, he sent

a truck load of steel cots, mattresses and blankets, as most would stay from Saturday morning to Sunday afternoon. To extend the

---

2. Other early instances of navigational error due to poor and incorrect maps did not turn out as well; some ended with both loss of life and aircraft. See Hinkle, 1970: 26–36 for an account of the disappearances of Lts. Frederick Waterhouse and Cecil H. Connally on a flight from Yuma, Arizona, to San Diego, California.

Refueling a Consolidated PT-3 from Brooks Field at Johnson's Ranch. Elmo
Johnson pumps gasoline from a fifty-gallon drum to Lt. Charles W. Deer-
wester, seated on the top wing, while Lt. John W. Egan observes below.
Robert Johnson, Elmo's father, is seated on the runway maintainer; Lee
Rackley, who worked for the Johnsons, stands on the wing behind the
engine. Courtesy Smithers Collection.

invitation to the flyers, the Johnson ranch and landing field was
described to a couple of flyers at each field and were told to tell
others. For the next ten or twelve years, the Johnson ranch was a
very popular place for the flyers. (Smithers Collection)

Smithers' and Foster's "Air Corps grapevine" soon extended to Air
Corps stations throughout Texas. When Lt. Charles W. Deerwester
and Lt. B. A. ("Bunk") Bridget, two avid hunters, heard of the new
field, they arranged an inspection flight on May 16, 1930, for a firsthand view of this much-discussed new facility. They made the flight
from Brooks Field in two Consolidated PT-1 primary trainers, the first
type of aircraft, other than DH-4s, to land at Johnson's Ranch. Both
men were intrigued with the Johnsons, the landing field, and the surrounding country. In recalling the pleasant memories of that first
flight, Deerwester said, "We gave such glowing accounts when we
came back from Johnson's Ranch that other pilots from Brooks Field
started flying down there on weekends" (interview, February 11, 1971).
Officially, most pilots who registered at Johnson's Ranch were either
on operational assignment combining cross-country navigation train-

ing with exercises in strange (unfamiliar) field landings, or using the field as a turn-around point while accumulating flying time. Turn-around time could also be scheduled on weekends, enabling the flight crews to combine duty with personal recreation. The recreational opportunities were varied: hunting, fishing, exploring Indian caves, burro rides across the Rio Grande to Mexico, and visiting with the Johnsons. Others came just to escape the military routine of the San Antonio–area bases, lounge in the cool, vine-covered patio, and enjoy Ada Johnson's home cooking.

Hunting—mainly quail and deer—became a high-priority activity. Gen. Hugh A. Parker, a hunter, close friend of the Johnsons, and a frequent visitor to the field, explained that guns were tools of the military profession, and the wide-open spaces of the ranch offered the airmen ample opportunity to enjoy firearms as a related hobby. Parker never passed up an opportunity to fly to Johnson's Ranch. "I used to like to shoot and you could just stand out in the front yard, and any direction you wanted to shoot, you could have at it. And I liked Elmo and I liked Mrs. Johnson" (interview, February 8, 1978).

The charm and hospitality of the Johnsons became the major attraction of this Airways Section landing field. Once on the ground at the ranch good fellowship prevailed, day-to-day military responsibilities were forgotten, and, most important of all, everyone disregarded military rank. For the most part, flights to Johnson's Ranch were for the good times and the Johnsons made them so. Elmo Johnson became the principal symbol of this remote Airways Station and, consequently, the field evolved totally unlike any other in the chain of Air Corps facilities. Johnson's Ranch existed within that vague twilight zone between military officialdom and casual individualism. Former Air Corps officers who frequented this facility attest to this fact, and Col. Arthur G. Fisher, Eighth Corps Area Air Officer, accorded official sanction to this rare military condition: "No rank will be observed at Johnson's Ranch, except for Elmo, who is always in charge." Later Johnson added modestly, "This included generals as well" (interview, June 3, 1966).

Remoteness, as well as the absence of a permanent staff, contributed to the relaxed atmosphere at the Big Bend airfield. In comparing this field with other similar facilities in the nation, Egan concluded, "There probably weren't any that were more primitive than this, further removed from civilization by distance. Certainly within the continental limits of the United States, this was about the only one" (interview, February 7, 1978).

For whatever purpose and within whatever operational framework flights were made to Johnson's Ranch, the presence of United States military aircraft operating along the Mexican border helped fulfill one of the field's primary purposes—a display of force to deter violations of the international boundary. By the 1930s, a large measure of the responsibility of protecting the local citizenry rested with the Air Corps' ability to deploy combat aircraft to the Big Bend airfield under emergency orders. Either by command decision or personal preference, this responsibility fell mainly to Lt. Thad Foster, who, according to Colonel Egan, could always "fix up an excuse to fly . . . [and] each flight had military purpose" (ibid.).

Foster's apparent eagerness for combat readiness is noted in a brief entry in the field register dated June 6, 1930: "C-1C Lt. T. V. Foster - Lt. Bone - Sgt. Hodges - Pvt. Wagstad 8CA." On this date, Foster, accompanied by personnel from the Eighth Corps Area headquarters, made a familiarization flight to Johnson's Ranch in a Douglas C-1C. This aircraft was the Air Corps' first cargo-personnel transport capable of deploying armed troops to a combat zone. In retrospect, Foster's flight in the C-1C did indeed have military purpose, because within the coming year this simulated operation would be repeated with more efficient equipment and under combat-ready conditions.[3]

The Douglas C-1C was the largest military aircraft to land at the new Airways Station. This aircraft was a modification of the historic DWC (Douglas World Cruiser) designed by the Douglas Aircraft Company for the 1924 around-the-world flight sponsored by the United States Air Service. The enclosed cabin contained seats for eight fully armed soldiers, while the pilot and crew chief sat in an open cockpit immediately behind the engine. Its design and performance data, however, reflect the elementary state of aerodynamics in the mid 1920s. The wooden-constructed, fabric-covered biplane was powered by a four-hundred-horsepower, water-cooled Liberty engine that gave the angular box-like airplane an actual flying speed of about seventy-five miles per hour. Gen. William L. Kennedy was less than enthusiastic about the air time he spent in the open cockpit of the C-1C. "I used to fly that son-of-a-bitch," he recalled. "Among its peculiarities was, if you left the front windshield on, the propeller blast buffeted the top wing and eventually tore off the fabric. So you would have to leave the front windshield off. Well, that was alright, because

---

3. Gen. William L. Kennedy identified the "Lt. Bone" on this flight as Lieutenant Norfleet Bone, who planned and supervised the landscaping of Randolph Field during its construction. The C-1C was used to bring back native desert plants for that purpose. This flight is dated June 6, 1930; Randolph Field was dedicated on June 20–21, 1930.

you flew with helmet and goggles anyway—until you ran into rain!" (interview, February 20, 1979).

Cockpit discomfort is what Gen. P. H. Robey remembered most graphically about the hours he loged in the Douglas C-1C. "You sat right behind the engine," he explained, "and the fuselage tapered noticeably from the point where you sat, to where the old Liberty engine was attached. Well, that placed your feet almost against the crankcase! After flying that thing for a while, talk about getting a hotfoot!" (interview, February 16, 1981). Kennedy conceded, however, that "actually it was a pretty good airplane for its purpose. You could land it short, you could take it off short, and it carried a lot." "And it scared the hell outta you every minute," interrupted Egan, who also flew the C-1C. "It was a damn box with a couple of wings, and a Liberty engine out front to drag it through the sky—by brute force!" Referring to the aircraft's slow speed, Kennedy added, "I've seen cars pass me as I flew along Texas highways" (interviews, February 20, 1979). For all of its shortcomings, the Douglas C-1C fulfilled the Air Corps' need to deploy armed troops via air. Of even greater importance, this vintage troop carrier/transport was the forecast of things to come.

During the decade of the twenties, America generated a growing interest in flying and, until the stock market crash of 1929, eager investors speculated heavily in aviation issues, many of dubious worth. (In 1930, ninety-four different types of private and commercial aircraft were being manufactured in the United States.) Economic depression, however, eliminated much of this investment capital and many firms succumbed to bankruptcy. During the early 1930s, however, military and air carrier orders kept the struggling aviation industry solvent. The Douglas Aircraft Company survived, and by the time Lieutenant Foster landed the C-1C at Johnson's Ranch, new airframe designs were already on the company's drawing boards—trainers, observation types, attack planes, and more transports. Although vastly different in design, the C-1C evolved as the first in a long series of Douglas transports, which, over the coming decades, gave added punch to the United States' aerial arsenal as well as revolutionized the airline industry. One version helped bring victory in World War II; another pioneered commercial air routes to every corner of the globe. The Air Corps designated the C-1C's distant cousin the C-47; the airlines' version evolved as the DC-3. Both became American household words.

The flight of the C-1C and the periodic landing of military aircraft at Johnson's Ranch portended far more than immediate security. The routine operation of the Air Corps planes in an area that for more than two decades had been the private stronghold of the United States Cavalry symbolized vast technological advances that were occurring throughout Western civilization. In industrial societies both man and

Johnson's Ranch became the setting for the Army Air Corps' early troop carrier operations. In May 1929, the Eighth Corps Area command dispatched two Douglas C1-Cs (above) to the Big Bend airfield, each carrying a squad of fully armed infantry soldiers. This was a visual warning to Escobar rebels reported in that area. On September 15, 1931, a Fokker Trimotor C7-A (below) landed eight infantrymen at Johnson's Ranch on the "Border Raid Mission." Courtesy Aerofax, Inc./Jos. Nieto Collection.

animal were surrendering responsibilities to the machine, and the proponents of the horse cavalry standing conservatively in the path of this techological ground swell could not long remain oblivious to the forces of change. To many observers, it was obvious that the cavalry's traditional beast of burden was doomed to be replaced by newer mechanical devices of war.

For almost two decades Camp Marfa (later Fort D. A. Russell) had served as headquarters for various cavalry units that maintained border security throughout the Big Bend country. This was a period of high drama along the Rio Grande. The 1914 rescue of the Mexican refugees from Ojinaga, the Glenn Springs raid, the Brite Ranch raid, the Nevill Ranch raid, Chico Cano's reign of terror, and, more recently, the Johnson Ranch raid all fell within this time frame. The cavalry's extended tenure in this region sustained the War Department's commitment to protect the Texas border. As the decade of the 1920s drew to a close, however, a new surveillance posture began to emerge. Some of the first ripples on a sea of military calm caught the attention of observers during the cavalry's annual field maneuvers. The more astute could foresee a ground swell in the offing.

The United States Cavalry had conducted these joint combat simulation exercises in the trans-Pecos region throughout the decade of the 1920s. Since the Air Corps observation squadrons (sometimes referred to as the "eyes of the infantry") functioned in accordance with the needs of the ground forces, these maneuvers brought the services together in combat exercises. The War Department accorded a high priority to the results of these maneuvers as, following World War I, the changing roles of both the mounted cavalry and the air service were being carefully reevaluated. According to photographer-journalist W. D. Smithers, who observed Texas military activities from 1916 to 1932, "many farsighted Cavalry officers realized by the end of World War I that the airplane was destined to change the entire concept of modern warfare, and the horse cavalry's role would be either terminated, or relegated to a vastly different assignment" (correspondence, July 5, 1979).

The sparsely populated stretches of the trans-Pecos region provided a favorable setting for the cavalry to demonstrate its continuing worth to the body military. The 1929 maneuvers, which began on October 8 and ended on October 26 with a division review, attracted wide attention. It was, however, the newspaper editors who articulated for an interested public the latent interservice conflict. With military aviation still in its teenage years, the verbiage for the most part stemmed from the past rather than forecasting the future. One editor, focusing on the competitive aspects of the maneuvers, informed his readers that

Lone DH-4 circles the cavalry encampment at Camp Marfa during the 1927 Air Corps/cavalry maneuver. The photograph seems to suggest that the future belongs to the Air Corps, at that time referred to only as the "eyes of the infantry." Courtesy Smithers Collection.

these maneuvers probably will include many tests of aircraft operating against mounted troops and armored cars. Pilots, aerial gunners, bombers and observers have not yet fully overcome the handicaps of fog, storms and darkness. These phenomena are obstacles to cavalry, as well, but in lesser degree. The two arms supplement each other strongly. . . . As the United States has large areas which almost or wholly lack highways, it should not further reduce its cavalry. (*San Antonio Express*, October 5, 1929)

Although the editor understood United States geography and climatology, he failed to anticipate the rapid pace of American technology. And while he based his arguments on previous data and current limitations, he apparently was unaware of other areas of disagreement that further strained the Air Corps–cavalry brotherhood: cavalry officers' decreasing opportunities for advancement and, most important, their traditionally lower pay scale. The latter point, plus the personal freedom the Air Corps officers enjoyed, ruffled many military feathers. This was not a recent development; interservice rivalry predates the 1929 maneuvers by at least a decade. An Air Service pilot who served with the 1920 Border Air Patrol wrote:

The cavalry officers never felt very friendly toward the flying officers since pilots were not under the same strict regulations as cavalrymen, and both in temperament and in type of work done, the differences were great. It was especially irksome to the cavalry officers that pilots received an additional 25 per cent of base pay for flying. This rankled worse when the additional pay went up to 50 per cent in 1920. (Hinkle, 1967: 24)

Declining opportunities for permanent commissions in the cavalry strained further the relationship between the mounted officers and those who flew the airplanes. Col. LeRoy Hudson, one of the Air Corps pilots who participated in the 1929 maneuvers, explained: "In those days, as the horse cavalry was being phased out and the air arm was growing, [and] the number of spaces [permanent officer commissions] allocated to the air was taken from them, it did cause a little friction, mostly conversation of air warfare versus cavalry tactics." Hudson then added in retrospect, "And, of course, how silly the War Department was not to see the wars could be fought without cavalry and horse drawn artillery" (correspondence, January 2, 1979).[4]

Faced with mounting evidence that the era of the old-time, tradition-bound, polo-playing cavalry officer was nearing an end, many of the far-sighted leaders accepted the challenge of change while continuing the struggle to maintain the high order of the service. In this effort, some of those who were destined to become leaders in the new mechanized cavalry gained the support and admiration of their Air Corps counterparts. Gen. William L. Kennedy, another Air Corps pilot who participated in these joint exercises, explained that he

worked with the cavalry and infantry and my own personal impression was that the cavalry was pretty good to work with, because they were in trouble and knew they were in trouble. *They would do things.* The infantry would sit on their butts for days at a time and nothing would happen in a maneuver of combat simula-

---

4. In the early 1930s, the Air Corps was still operating under principles laid down by the War Department General Staff in 1920. This created much dissatisfaction among Air Corps officers at command level since all Air Corps combat units were under the command of ground officers at division, corps, army, and GHQ levels. During the 1920s and early 1930s, this led to considerable conflict "between air and ground officers over the composition, organization, and control of military aviation." On March 1, 1935, this matter was largely resolved with the creation of the General Headquarters Air Force (GHQAF), which placed the air defense and striking forces under the control of air force commanders (Maurer Maurer, 1961: 4–6).

tion. I liked working with the cavalry . . . because they were in there trying. And the best men out of the bunch ended up in the armored cavalry and one of them was named [Gen. George S.] Patton, [Jr.]. (Interview, February 20, 1979)

Although the war games conducted near Fort Bliss helped sharpen the participants' tactical proficiency, they achieved little toward establishing the nation's ultimate defense posture. Time, technology, and world affairs would help make those determinations. A half-century later Colonel Hudson could not even recall who won or lost the mock encounter. He did remember, however, "the unofficial critique which took place after it was over. This was held, of course, in a foreign neutral country, Juárez, Mexico, as prohibition held sway in the U.S." (correspondence, January 2, 1979).

The unofficial critique in Juárez, followed by the cavalry review and polo tournament, ended the 1929 maneuvers. Other joint war games would be held in succeeding years. But for the present, military protection of the Texas–Mexican border remained the joint responsibility of the two branches of the service, with coordinate points located at Johnson's Ranch and Camp Marfa. The establishment of the Big Bend airfield, however, reflected a new philosophy in border security. The original dirt strip that stretched northwest from the Johnson trading post afforded the Air Corps immediate access to the region. And when Lieutenant Foster's DH-4 first touched the crusty soil of Elmo Johnson's ranchland, it symbolized the beginning of the end of the horse cavalry's tenure in the Big Bend of Texas and the growing importance of air power in the United States military strategy.

Obstructions to change within the defense establishment are many and represent issues that span the military spectrum from the days of Caesar and the Roman Legions to the twentieth-century debates over aircraft carriers, strategic bombers, and intercontinental ballistic missiles. Ultimately internal grievances succumb in the path of material progress, old bastions fall, and the process of change is relegated to the historian, perpetuator of military tradition.

In the early 1930s, the strategic roles of man, the horse, and the machine—both aerial and earthbound—were in flux. The resolution of conflict within this military drama would, however, be resolved in the not-so-distant future. Science and technology, not boards and commissions, would dictate the ending. And while presidents, prime ministers, dictators, and chiefs of staff debated the issue, one of the closing acts in this world drama would be played out in the remote vastness of the Big Bend of Texas. Elmo Johnson and his Air Corps friends would be present for the final curtain.

"... my gosh, war's broken out"

CHAPTER 7

The decade of the 1930s marks a vast watershed in the span of American history. Beginning with the Great Depression and ending with the United States' entry into World War II, this ten-year time frame bisects two periods of the American experience that, from any perspective, differ diametrically. In retrospect, the far side of the plateau reveals an uncomplicated American society, primarily agrarian, largely static, whose social, political, and economic ideals were rooted firmly in the past. As the decade progressed, however, a new image of American society began to evolve: old power structures were crumbling, traditional behavioral patterns were disappearing, and the American economy became the focal point of national concern. But amid the suffering and hardship that came in the Depression's wake, there appeared a faint light at the end of the tunnel, forecasting a new society totally unlike any that had gone before.

During the watershed years some of the more radical developments were occurring in the fields of science and technology. These would lead America to the threshold of some of the most dramatic changes in world history. Aviation was part of this technological growth and its role in the evolving scheme of things would dramatically alter American society during the decade of the 1930s. Industry leaders would pioneer new methods of travel, transportation, and communication, while military strategists would push the airplane's destructive capacity to levels inconceivable when the decade began. Johnson's Ranch, nestled deep in the Big Bend of Texas, far removed from the seat of the decision makers, would ultimately reflect the changes that would project America well beyond this great historical divide.

The slightly more than a decade that separated the end of World War I and the dedication of the Big Bend airfield was a period of antimilitarism, dominated by an isolationist psychology that rejected both the Treaty of Versailles and the League of Nations. During this period Congress held military appropriations to a bare minimum, and the Air Service, administrative stepchild of the United States Army, suffered the most. It was, temporarily, the nation's undernourished combat arm fighting for status, role, and scope. Hence the continued but somewhat limited use of the DH-4s, described by air power's flamboyant and volatile crusader Gen. William ("Billy") Mitchell as "neither fish, flesh nor foul as they are neither attack, bombardment, pursuit, nor observation planes" (Glines, 1963: 120). The failure of the Air Corps to alter the DH (De Havilland) designation substantiates further the articulate general's pronouncement.

All was not bleak, however, as the Air Corps Act of 1926 created an administrative framework for establishing a viable air force. Two key facets of this legislation provided for an expanded research and development program as well as a continuity of orders of new military aircraft with a standard rate of replacement. Hence the ultimate demise of the venerable DH-4.

Although the last of the DH-4s were not retired from service until 1932, Lt. Thad Foster registered the last landing of this World War I relic at Johnson's Ranch on July 19, 1930, marking the end of an era. Significantly, this flight was made in the company of a Miss Clarissa Ryan, a decision on the part of the control officer that may or may not have added to the historical significance of the occasion.

Foster returned to Johnson's Ranch on August 16, flying a Douglas O-2H. This was the first in a long series of Douglas two-place observation and training biplanes that played an important role in the Air Corps cadet program as well as in the ultimate development of the independent concept of tactical reconnaissance aviation. (The O designates the aircraft as an observation type that functioned primarily in conjunction with ground force operations.) Powered with a 450-horsepower water-cooled Liberty engine, the O-2H could attain a top speed of 137 miles per hour and cruised comfortably between 100 and 110 miles per hour. Air Corps pilots regarded the DH-4's replacement as an extremely sturdy, dependable, and stable aircraft. Gen. Robert S. Macrum remembered that after receiving instructions in primary trainers "we moved on to DH-4s, a heavier type of aircraft." During the final one-third of the course, "the O-2Hs started coming in. Beautiful, beautiful, Douglas aircraft. Oh, they were beauties after flying the DH-4s" (undated interview, 1979).[1] Within the space of a few months, this new

---

1. The success of the O-2H led to the Air Corps' subsequent purchase of two additional versions of this Douglas biplane. The O-38 employed the same

series of aircraft gave the Air Corps increased performance capability over the DH-4s and extended the safety factor manyfold. This was a prime consideration for the pilots negotiating the unpopulated desert terrain along the Rio Grande that led to Johnson's Ranch.

During the second full year of operation, July 6, 1930, to July 5, 1931, the Air Corps cleared twenty-five flights to Johnson's Ranch, representing a total of forty-eight personnel (including Miss Ryan). Foster remained the station's most frequent visitor, registering seventeen times during the year. One noticeable change is noted in the control officer's entries for this period. High-ranking officers frequently accompanied Foster to the Big Bend airfield, many listing their rank as either major or colonel, a significant fact in the Air Corps administrative structure during the 1930s.

The Air Commerce Act of 1926, which renamed the Air Service the United States Army Air Corps, failed to provide a service academy at which to train command officers. Most senior assignments in the Air Corps, therefore, went to officers transferring from other branches of the services; practically none were trained pilots. One of these interservice transferees was Lt. Col. Arthur G. Fisher, Eighth Corps Area air officer, whose command included Johnson's Ranch. Fisher, a "balloonatic" who had previously served with the army's lighter-than-air service, was not a trained pilot. This was a shortcoming that spawned varying degrees of animosity among the junior Air Corps flying officers. Gen. William L. Kennedy, a second lieutenant in 1930, remembered Fisher as "an oldtimer and a very courtly old man." He was, however, "not a pilot. At that time, of course, pilots took a dim view of anyone in the Air Corps who wasn't a pilot. But, nevertheless, he was a pretty nice old guy" (interview, September 11, 1979).

The responsibility for flying Fisher on inspection trips throughout the Corps Area fell to his assistant, Lieutenant Foster. Fisher, like Foster, was obviously committed to his assignment, because during the year the two officers registered at the Big Bend airfield on six different occasions. Fisher first inspected Johnson's Ranch on November 6, 1930, and was the highest-ranking Air Corps officer to visit the facility up to that time. On February 6, 1931, Foster piloted a Major Coker to Johnson's Ranch on an inspection trip and returned with him on March 8, flying a Douglas BT-2B, the first aircraft of this type to land

---

airframe, but the 525-horsepower air-cooled radial engine increased the top speed to 149 miles per hour. The Douglas BT-2 series (basic trainers), equipped specifically for dual flight instruction, cruised at 113 miles per hour with a 450-horsepower radial engine.

at the airfield. Coker, another nonflying officer, was attached to the Eighth Corps Area Headquarters at Fort Sam Houston and subsequently became the Corps Area air officer, succeeding Colonel Fisher.

Foster's "aerial taxi service" to Johnson's Ranch was not limited to girl friends and ranking Air Corps officers. On August 6, 1930, he flew Fort D. A. Russell post commander Col. John S. Fair to the airfield, and on October 20 he provided the same service to Col. Joseph C. King, another Fort D. A. Russell cavalry officer. On December 10, while on a trip from Dodd Field to Johnson's Ranch, Foster landed his BT-2B on a "flat place" near Sonora, Texas, picked up Elmo Johnson, who was visiting his family, and returned him to the ranch.

Although the Air Corps operational procedures forbade flying civilian personnel in military aircraft without official sanction, Elmo Johnson occupied a unique niche in the administrative structure of the Southwestern Airways Section. He owned the landing field, was personally responsible for its maintenance, and was a close friend of the air officer, the control officer, and most of the Air Corps pilots who regularly used the Airways Station. In addition to his personal relationships, Johnson was accredited to the G-2 Army Intelligence Section of the Eighth Corps Area headquarters at Fort Sam Houston. Kennedy explained Johnson's working relationship with the Corps Area as follows:

Johnson was accredited to the Army Intelligence. They called it CID, Counter Intelligence Department, back then, but actually what he was was a border watch. . . . They needed responsible men along the border who could report anything peculiar. In the wake of the bandit era, the decision to maintain constant surveillance was valid. That explains Johnson and our association with him. He could fly in our airplanes because of this accreditation. (Interview, January 19, 1978).

Johnson, if not modest, was more succinct in describing his responsibility to the Corps Area headquarters: "It was up to me to keep the Army informed on what was going on along the border" (Cox, 1970: 63).

Col. LeRoy Hudson, a former Southwestern Airways Section air officer, described the Big Bend facility as G-2's "isolated window into the happenings along the border." He remembered "it was remarkable that for an area so isolated, there was so much traffic between Mexico and the U.S." (correspondence, September 20, 1978). Johnson's qualifications as the Air Corps' premier border watch attracted the attention of Gen. P. H. Robey, a frequent visitor to the field in the early

1930s. He explained that Johnson was "well known on both sides of the border, highly respected, and of course, privy to all the goings-on in that entire area" (correspondence, November 14, 1978). It was, however, the trading post that actually provided the "isolated window" to which army intelligence accorded priority interest. The constant flow of Mexican nationals to the riverside emporium afforded Johnson a prime source of information on the internal affairs of that section of northern Mexico. Gen. Hugh A. Parker, a close friend of the Johnsons, explained: "There was usually a crowd over there. . . . Several Mexican families would cross the river to come to Elmo's during a twenty-four hour period" (interview, February 8, 1978).

Elmo Johnson's information-gathering chores along the Mexican border were part of a broad intelligence program being pursued throughout the Eighth Corps Area. Maj. Robert J. Halpin, assistant chief of staff for Military Intelligence, directed this effort. On August 9, 1930, he forwarded the following confidential memorandum to a Captain Watkins, the Fort D. A. Russell intelligence officer:

> Owing to the uncertain situation south of the Rio Grande, it is requested that you be especially alert to obtain and forward any information relative to new developments. Please bring this to the attention of your commanding officer. . . . Frequently small items and even rumors may fit into our blotter of information and be of real value. (RG 394)

Command-level concern for border security prevailed well into the following year. On January 12, 1931, Maj. Gen. Edwin B. Winans, Corps Area commander, instructed Major Halpin, "by authority of the Secretary of War," to visit personally twenty-four border sites and installations "in order to carry out the confidential instructions of the Corps Area Commander." Major Halpin's itinerary encompassed the entire border area extending from Port Isabel, Texas, to Camp Stephen D. Little at Nogales, Arizona. Of the nineteen Texas installations he visited, nine were in the Big Bend area. These included Fort D. A. Russell, Presidio, Boquillas, and Johnson's Ranch.

The local information Johnson gathered at the trading post appeared consistent with the broader scope of international relations. Conditions along the Texas-Mexican border had become uncharacteristically quiet. For more than two years, reports of border incidents were noticeably missing from the newspapers, and the various political factions, so long active within the Mexican government, began resolving their differences with debate instead of bullets.

In the welcomed calm that followed the raid on his trading post,

Johnson recalled later that he was surprised when, in early September 1931, he heard rumors of impending border trouble from some of his Big Bend friends. They had heard that on September 16, a date for traditional Mexican independence day celebrations, Mexican bandits were planning a series of raids on the ranches and trading posts along the Rio Grande.

Since he was unable to verify these reports from his usual sources, Johnson decided the rumors were false. Reflecting on the event years later, he remembered he "thought it was a fake, but one morning a couple of cars of Rangers and deputies with guns sticking out of the windows roared up to my trading post." The officers repeated the reports of the impending attack, adding that "all the women are being moved out. You'd better get your wife away." Johnson told the lawmen that "my wife was almost as good with a gun as I was and if they were going to raid, we'd play with 'em" (Cox, 1970: 63).[2]

Following the appearance of the Rangers, Johnson "still didn't think much of it, but just routinely I sent a message about it to the Army" (ibid.). About noon the same day, September 12, 1931, Lt. Sigma Gilkey and Maj. Keith Simpson arrived at Johnson's Ranch in a Douglas BT-2 from Brooks Field. Gilkey was attached to the Twelfth Observation Squadron at Brooks Field, while Simpson, a physician, served as flight surgeon with the Aviation School of Medicine at the same base. Both were frequent visitors to the Big Bend airfield. The two officers joined the Johnsons for lunch, during which they discussed the rumored raid. The conversation ended with Johnson suggesting that, in addition to his telegraph report, it might be wise for them to contact the Corps Area headquarters about the matter when they returned to their base later that afternoon. Significantly, news of the impending raid must have been circulated in the area subsequent to September 8, because Lieutenant Foster and Colonel Fisher visited the airfield on that date and Johnson made no mention of his discussing the matter with them.

Johnson and his wife, still ignoring the Rangers' warning, left the trading post on the morning of September 15 for a day-long visit to

---

2. In this publication Johnson refers to the date of the proposed raid as May 5, Cinco de Mayo, another traditional Mexican holiday. Although Johnson gave the author the same date, he obviously erred. Newspaper accounts and the recollections of Air Corps officers who participated in the subsequent reconnaissance missions confirm the September 16 date, Diez y Seis de Septiembre. Texas Ranger records for the pre-1936 period are contained in the adjutant general's and governor's papers in the Texas State Archives. These sparse records contain no information on impending border uprisings.

Study (pronounced Stoody) Butte. The thirty-five-mile trip to this quicksilver mining village was over some of the worst roads in the country, and they did not return home until late in the afternoon. Johnson long remembered his surprise as they approached the airfield and saw the line of parked aircraft guarded by armed soldiers. "It looked like the whole Army had moved to my trading post. 'My gosh, war's broken out,' I told my wife" (ibid.). As Johnson parked his truck beside the runway, the first person he recognized was Colonel Fisher, who shouted to him, "Elmo, where's this war you want us to fight?" (interview, June 3, 1966).

The field register entry, cited as the "Border Raid Mission," indicates Colonel Fisher had indeed delivered a potent aerial and ground strike force capable of fighting any war likely to erupt along the Rio Grande. Fisher and Foster led the flight in a Douglas BT-2B, the only noncombat aircraft on the mission. Capt. W. S. ("Pop") Gravely, operations officer with the Twelfth Observation Squadron at Dodd Field, and Lt. John W. Egan, attached to the same unit, flew Thomas-Morse O-19Es, "with machine guns on 'em loaded up with all the ammunition we could carry" (interview, February 7, 1978). Gravely's observer-gunner was a Sergeant Jones, attached to the Ninth Infantry Regiment at Fort Sam Houston, while a Private Tynebury from the same unit occupied Egan's rear cockpit. The O-19E carried two .30 caliber machine guns, one synchronized to fire forward through the propeller and the other mounted on a swivel base in the rear cockpit.

Fisher's biggest gun was a Fokker Trimotor C-7A, the Air Corps' latest and most advanced cargo-personnel transport. Piloted by Lt. Ike Ott, this high-wing monoplane had a wing span of 72 feet 10¾ inches, was powered by three 330-horsepower Wright "Cyclone" air-cooled radial engines, and carried a useful load of almost 4,800 pounds at a cruising speed of 110 miles per hour. This was the Air Corps' first multi-engine cargo-personnel transport and reflected major advances in aircraft design. The C-7A performed with improved speed, capacity, range, and dependability over the Douglas C-1C, which Foster had previously flown to Johnson's Ranch.

Lieutenant Ott's passengers on the C-7A included his mechanic (a Corporal Tyler), plus a squad of infantrymen—two sergeants and six privates—also from the Ninth Infantry Regiment. The infantrymen were armed with the standard Springfield rifles, and one was assigned the new Browning .30 caliber automatic rifle, a highly revolutionary weapon at that time. Egan explained that "most infantry squads had one automatic rifle. . . . The rest of 'em were still carrying the Springfield" (ibid.).

This was Egan's first trip to Johnson's Ranch and, although Colonel

Fisher was the ranking officer on the mission, he credits Lieutenant Foster with organizing and administering the attack force. According to Egan, Foster arranged for the aircraft and the squad of infantrymen that accompanied them on the mission. "We flew down there," Egan explained, "landed at the field, had a guard on our aircraft, and camped in Johnson's patio. We were there about three days and made several flights up and down the river" (ibid.).

In retrospect, it appears that all information pertaining to the impending raid emanated from either law enforcement officers at Marfa, Texas, or from the First Cavalry Regiment headquarters at Fort D. A. Russell.[3] More significantly, the presence of combat aircraft in the Big Bend on the eve of the alleged September 16 encounter resulted entirely from Elmo Johnson's communications with the Corps Area headquarters. It soon became obvious that the operation lacked interservice coordination. And, judging from the First Cavalry Regiment commander's subsequent reaction, the presence of the airborne attack force at Johnson's Ranch was uninvited, unwanted, and totally unexpected.

Shortly after the air unit arrived at the Big Bend airfield, "word came down from Fort D. A. Russell at Marfa for whoever was in command of the troops to report at once," Johnson remembered. "That was pretty funny, because it so happened that the colonel who was in charge of the troops at my trading post [Colonel Fisher] outranked the commander at the Fort [Col. William A. Austin]." Fisher accepted the mistaken reference to rank good-naturedly, saying "he'd take his whole command to the fort and have every man shake hands with the commander there" (Cox, 1970: 62).

On the morning of September 16, the date of the alleged raids, the flight of four airplanes took off from Johnson's Ranch and began a low-altitude patrol of the Rio Grande. Foster and Fisher led the mission on several flights up and down the river but, according to Egan, they "never did see any mounted men, and if they were there they heard us. We made sure that we were flying low enough and making enough noise so that they would hear us and know we were there." The group patrolled the Rio Grande between Boquillas and Presidio, at which point they landed and Foster crossed the river to Ojinaga to confer

---

3. On January 1, 1930, Camp Marfa was renamed Fort D. A. Russell. At the suggestion of President Herbert C. Hoover, Fort D. A. Russell at Cheyenne, Wyoming, was renamed Fort Francis F. Warren, honoring the late senator from Wyoming. Hoover also suggested that Secretary of War Patrick J. Hurley select some other post that would commemorate the name of Gen. D. A. Russell, veteran of the Plains Indians Wars and the Civil War.

with some Mexican officials there. They in turn "sent a couple of *rurales* [Mexican federal border guards] down the river and they showed up at Johnson's about two days later" Egan explained. "We [later] met them at Johnson's and they hadn't seen any bandits" (interview, February 7, 1978).

Taking off from the Presidio field, the flight continued to Fort D. A. Russell, where Fisher reported to the insistent cavalry colonel. Egan does not recall their meeting, but Johnson, who accompanied the flight, remembered that Fisher did not go through with the mock handshaking ceremony. Their meeting, if not cordial, was at least perfunctory, and whatever this cavalry colonel hoped to accomplish by ordering the Air Corps detachment to the cavalry post was never forthcoming. Newspaper correspondents sent to Marfa to report the cavalry's response to the border emergency were fascinated with the speed, firepower, and flexibility of the Air Corps unit and seized upon an unexpected facet of the story. Johnson exhibited his military partisanship as he recalled the event with great relish: "The newspapers got hold of the thing and wrote a story on how much faster the Air Corps could cover West Texas than the cavalry. It started a movement to abandon the fort" (Cox, 1970: 62). Johnson was partially correct, but for the wrong reasons.

The future status of Fort D. A. Russell had been under consideration long before the "Border Raid Mission." According to W. D. Smithers, who had reported Big Bend military activities for more than a decade, the United States War Department originally planned to abandon the post in 1920 when the river outposts were closed. With border incursions continuing well into the 1920s, the cavalry post remained active for another decade. This was a great economic windfall for Marfa. The wholesale and retail trade generated by the post stimulated business activity throughout the region, adding a financial buffer to the local economy heretofore based largely on ranching and mining.

Following two years of relative calm along the Rio Grande, reports of impending bandit raids again focused the community's attention on Fort D. A. Russell. Yet the entire sequence of events lacked credibility. Elmo Johnson as well as the press representatives sent to cover the border emergency began to sense that somehow all the advance publicity accorded the purported bandit raids, the colonel's insistence that he coordinate the operation from the cavalry post, and the failure of the bandits to appear on the appointed date were all tied in with the uncertain future of Fort D. A. Russell.

The Associated Press coverage of the military alert focused on the presence of armed military aircraft, the refusal of unidentified officials to confirm reports of the impending raids, and the veil of secrecy surrounding the emergency preparations. Significantly, no mention is made of the mounted cavalry. The *San Angelo Standard-Times* carried the wire service story with a front-page headline. While this account judiciously traces the chronology of events, the story, nevertheless, is composed with a subtle undertone of guarded suspicion and disbelief. Highlights of the account follow:

> Army planes from San Antonio, maneuvering in this section the past two days and other unverified reports of projected bandit raids have prompted a few ranchers in isolated sections along the border to remove their families to population centers. Whether grounds for reports of possible border disturbances existed could not be learned here due to refusal of officials to comment. . . . Reports reaching here from official quarters in Mexico City discredited reports of any raid in this area. . . .
>
> Peace officers here maintain the strictest secrecy, but it is known that both counties [Brewster and Presidio] are being guarded against possible raids of Mexican bandits from across the border. . . . Presence of army planes with machine guns mounted, and a transport plane with soldiers armed with automatic rifles has been attributed to the unrest along the Rio Grande caused by the expected raids. (*San Antonio Standard-Times*, September 20, 1931)

The concern of local Marfa citizens for the future of the cavalry post is noted in the news service coverage: "Citizens recently have been urging government officials to retain Fort D. A. Russell, an army cavalry post near here, slated for abandonment under the present plans of the war department" (ibid.)

The *Standard-Times* story also contains an official Air Corps statement issued from San Antonio that omits any reference to an emergency. Gen. Edwin B. Winans, Eighth Corps Area commander, described the flight of armed military airplanes as "a routine training mission such as must be carried out at periodical intervals." Then, apparently addressing himself to Elmo Johnson's intelligence dispatches, the general added that "conditions along the border are quiet in so far as reports coming to Corps headquarters disclose" (ibid.).[4]

But all was not quiet in the Marfa business community, nor had it

---

4. A search of Record Groups 18, 94, 92, and 394 uncovered no data pertinent to the reported raid, the Air Corps response, or the First Cavalry Regiment's subsequent surveillance of the Mexican border. Records of Fort D. A. Russell (and Camp Marfa) included in RG 394 though obviously incomplete contain mainly records of routine post operations—transfers, travel authorizations, arrival of new recruits, personnel references, and results of court-martial proceedings.

been since the previous May, when news reached the Big Bend that the War Department had listed Fort D. A. Russell for abandonment. Coming in the wake of recent expansion and development programs at the cavalry post, the announcement was totally unexpected. President Calvin Coolidge had signed the Hudspeth-Sheppard bill on January 25, 1927, appropriating $27,000 to purchase land for making the fort a permanent cavalry post, and three years later Congressman Claude Hudspeth secured an additional appropriation. In August 1930, construction began on four new sets of officers quarters and ten new sets of noncommissioned officers quarters. In addition, the army installed a new high-frequency radio transmitter and receiver in the post's communication center.

The increased trade the cavalry post generated in the Big Bend area created a false sense of economic security; Marfa greeted 1931 on an optimistic upbeat. Responding to an eastern journalist's prediction of economic doom for the nation, the editor of the *Big Bend Sentinel*, alluding to the resiliency of the West Texas economy, concluded, "There is little doubt that this gentleman overdrew the picture" (Earney, 1977: 3). Time, however, proved the editor wrong; the Depression ultimately brought hard times to Marfa. In early 1931 job opportunities began to disappear, salaries dropped, and the economic stagnation that already strangled business activity in the North and East spread throughout the Southwest. Bankruptcy and business failures dominated the 1931 Texas economic news (*Texas Business Review*, February 1931: 2).

Business outpaces government in its response to economic decline. With the federal government's intrusive role in business and economic reform still a presidential administration away, President Hoover turned first to the federal budget in an effort to slow the nation's economic decline. During a period of peace dominated by isolationist philosophy, the president's target for retrenchment was predictable—the military. On May 21, 1931, the War Department announced that President Hoover was meeting with his chief of staff, Gen. Douglas MacArthur, and the executive officers of the army to develop a plan for disposing of all unnecessary army posts and concentrating these troops at central points for a more united command. After several days of study the committee issued a list of fifty-three army camps slated for abandonment, five of which were in Texas, and included Fort D. A. Russell.[5] The news reached most Marfa citizens on May 28 via the *Big Bend Sentinel's* headlines:

---

5. War Department news release, "Army Posts to Be Abandoned," Sterling Papers: *Big Bend Sentinel*, May 28, 1931.

[Big] Bend to Fight to Retain Fort Russell; Flood of Protests Greet War Department Recommendation That Fort D. A. Russell Be Abandoned—Big Bend Region Arranges For Campaign to Retain Post at Marfa—See Severe Set-Back to Progress of Section if Troops Removed. (*Big Bend Sentinel*, May 28, 1931)

Significantly, the initial response focused on the "severe setback to progress"; concern for whatever protection the First Cavalry Regiment afforded was momentarily forgotten. The Marfa Chamber of Commerce coordinated the protest campaign as ranchers, businessmen, city officials, service clubs, civic organizations, and representative citizens throughout the area flooded Washington with telegrams. Texas senators Tom Connally and Morris Sheppard, Texas congressmen Harry Wurzbach and John Nance Garner, and Secretary of War Patrick J. Hurley were the key recipients. Meanwhile, a Big Bend delegation headed by L. C. Brite, a nationally prominent rancher, met with some El Paso businessmen to enlist their support "in convincing the War Department of the necessity of retaining a post in this immediate section." The El Paso contingent included Haymon Krupp, whose wholesale firm maintained retail outlets in the Big Bend area. Simultaneously, the Marfa Chamber of Commerce launched a fundraising campaign to carry on the fight to keep the cavalry post at Marfa (ibid.).

At the outset of the campaign, economic self-interest emerged as the key issue articulated by the Marfa business and ranching community. With the entire nation locked in the throes of economic depression, however, this posture lacked argumentative impact. Apparently overlooked in the immediate aftermath of the War Department's announcement was the reason for establishing the post in the first place—to protect citizens from Mexican bandits. To many living in close proximity to the Mexican border as well as those in high political and military circles, the Escobar Syndrome remained a compelling force.

The ensuing barrage of letters and telegrams are uniform in their argument, referring only to fear of bandits and citing Fort D. A. Russell as the last bulwark standing between the region and the raiding hordes from south of the border. By 1931 it was an outdated argument, but the Big Bend citizens gave it their best shot.

Telegram from R. E. Petross, Marfa Chamber of Commerce, to Texas governor Ross S. Sterling, May 21, 1931, 11:05 P.M. (my punctuation):

War Department recommendation abandonment Fort D. A. Russell must be vigorously and diplomatically contested. Reasons for establishment this post were never more real than at present. Economic and political conditions serious in Mexico. We of the Big Bend are exposed to raids and depredations, which without military show and protection, jeopardize life and property. (Sterling Papers).

Letter from Harry V. Fishers, commander of the American Legion Big Bend Post, No. 79, Alpine, Texas, to Secretary of War Patrick J. Hurley, May 27, 1931:

> . . . the presence of cavalry at Fort D. A. Russell is all that prevents frequent raids and robberies from Mexico on our frontier ranches, irrigated farms and commissaries. If troops are moved, we anticipate a series of raids, robberies and murders almost immediately—just as soon as the riffraff that infest the border learn that our isolated ranches have no further military protection. . . .
> These are cold facts. They are being handed to you by men like yourself who have already demonstrated their willingness to fight for their country. We do not mince words. (Ibid.)

Telegram from Texas governor Ross S. Sterling to Secretary of War Patrick J. Hurley, May 27, 1931:

> Citizens along Texas-Mexican frontier are greatly disturbed that forts Ringgold [near Rio Grande City in Starr County] and D. A. Russell are to be abandoned. If such action is contemplated, I hope you will reconsider for reason that people of Big Bend section along Rio Grande fear that abandonment of these posts would reopen that region to a reign of border banditry. (Ibid.)

Telegram from L. C. Brite, Big Bend rancher, businessman, and philanthropist, to Governor Ross S. Sterling, Texas senators Tom Connally and Morris Sheppard, Texas congressmen Harry Wurzbach and John Nance Garner, and Secretary of War Patrick J. Hurley, May 28, 1931:

> Am most emphatically in favor of retaining the post at Fort D. A. Russell. My ranch raided by Mexican bandits in December 1917 [garbled] as a result five men were killed [garbled] my ranch horses driven off and store raided. Loss amounting to several thousand dollars. The surrounding country was thrown into a state of terror. We appeal for continued protection. (Ibid.)

The secretary of war's brief response of May 29, 1931 to Governor Sterling's telegram discredited the need for protection, explaining, "These posts were garrisoned at a time when it was considered necessary that troops be stationed thereat, but it is believed that conditions have changed to such an extent that it is no longer necessary to maintain garrisons at those places" (ibid.).

The campaign to maintain the cavalry post yielded a measure of success. The week following his response to the governor's telegram, Secretary of War Hurley scheduled a brief stop at the Marfa airport while en route to Fort Bliss. Arriving in a Ford Trimotor transport plane, the secretary allocated only fifteen minutes for his Big Bend fact-finding mission. A select welcoming committee made up of ranchers, businessmen, and civic leaders greeted the secretary, who explained that he wanted firsthand information "concerning the necessity of maintaining a military post in the Big Bend section." Predictably, the committee avoided the economic issue, elaborating instead on Governor Sterling's doomsday prediction. Meade Wilson, president of the Marfa State Bank and owner of extensive ranch lands near the Mexican border, gave the issue an international scope. He claimed that maintaining the cavalry post "would prove an outstanding feature in cementing a better understanding with the sister republic to the south." According to Wilson, military officers from Chihuahua, Mexico, "had expressed a keen desire that U.S. military posts along the border be retained," as this would aid their forces in suppressing outbreaks of banditry and uprisings. There is no indication, however, that representatives from "the sister republic to the south" were invited to meet the secretary of war (*Big Bend Sentinel*, June 11, 1931).

If the secretary of war's brief visit to Marfa gave the Big Bend citizens cause for optimism, it was, in all probability, a political ploy to soothe the Texas-Washington delegation while the administration gathered support for its emergency policies. The secretary of war, and more especially the Hoover administration, faced issues of far greater magnitude than the future of a Big Bend cavalry post. For the first time in history all aspects of the national interest appeared in jeopardy, and most critical of all was the economy. In the face of a national emergency, the Marfa Chamber of Commerce's campaign to maintain the cavalry post was running counter to matters of national concern. Survival was the cry of the 1930s, and reduced military spending appeared to be a logical step toward that goal.

The crusade to keep Fort D. A. Russell proceeded unchecked throughout the summer of 1931, but chances for success appeared to be dwindling. On June 25, Congressman Claude Hudspeth met with Chief of Staff Douglas MacArthur to discuss the matter. The general explained that while the fort would ultimately be abandoned, no troops would be removed before the autumn of 1932 or possibly not until early 1933.

Cone Johnson, a member of the Texas Highway Commission who had achieved national political prominence, added his support to the campaign to save the fort. On August 21, 1931, he telegraphed his appeal to Secretary of War Hurley, who forwarded the communication to his chief of staff for a response. Four days later General MacArthur wrote Johnson, explaining that it was imperative that certain small isolated stations such as Fort D. A. Russell be abandoned. The general then injected an interservice issue into the controversy that had profound overtones: the Air Corps was being expanded at the cavalry's expense. MacArthur pointed out that reductions in budgetary consideration for the cavalry stemmed from the necessity of providing in excess of 1,200 men for the Air Corps annually, resulting in corresponding reductions in other branches of the army. For an officer steeped in American military tradition, crediting the growth of the Army Air Corps for the consequent decline in the ground forces undoubtedly ran counter to MacArthur's ground-oriented military philosophy. In explaining this to Johnson, the chief of staff was forecasting changes in the military establishment that would reach fruition during the decade of the 1930s.

Senator Morris Sheppard appears to have made a last-ditch appeal to the War Department on behalf of his West Texas constituency. His response, which came from Acting Chief of Staff Gen. George Van Horn Moseley, virtually closed the door to any further appeal to the War Department. After explaining the economic benefits of the army's relocation program then under study, he step by step destroyed the argument base so carefully developed by Big Bend citizens. Focusing on the military's expanding technology and the internal stability of the Mexican government, the general explained that

the development of motorized and mechanized units including armored cars, of our highway systems, and other modern means of communication permit rapid movements of troops from central locations and better protection of the border as a whole. Our Air Corps provides an increasingly efficient means of patrolling this area.

These changed conditions, together with the increasing stability of the Mexican government, are rendering unnecessary and undesirable the present dispersion of our forces in small garrisons. (Ibid., September 24, 1931)

Briefly, the general informed the senator from Texas that stable conditions along the Mexican border made Fort D. A. Russell "unnecessary and undesirable," and if trouble should arise either the Air Corps
or mechanized ground units—not the horse cavalry—would come to
the rescue. But, probably unknown to either of the generals, the decision to close Fort D. A. Russell would shift the Big Bend military spotlight from the cavalry post to Johnson's Ranch, where a cadre of young
Air Corps officers awaited the opportunity to demonstrate the growing
efficiency of air power.

When General Moseley dictated his letter to Senator Sheppard on
September 5, 1931, the Diez y Seis de Septiembre was only eleven
days away. The citizens of Marfa must have realized by then that
nothing short of a miracle could save the fort.

"... farewell to the horse"

CHAPTER 8

When the planes of the "Border Raid Mission" reached Fort D. A. Russell on the morning of September 16, 1931, the campaign to save the fort had been under way almost four months. During this time area civic leaders had enlisted the support of virtually everyone in the Big Bend, Austin, El Paso, and Washington who they believed could help further their cause. But all had come to naught; time appeared to be running out for the fort as well as the horse cavalry. The signs were unmistakable.

In marked contrast to the news coverage accorded the military alert by other Texas newspapers, the *Big Bend Sentinel* enforced a total news blackout of the event. The change in policy was dramatic indeed. Since May 28, 1931, reports on the campaign to save Fort D. A. Russell had dominated the *Sentinel*'s news coverage. However, the September 17 issue, published the day following the alleged raids, omitted any mention of the cavalry post, the "Border Raid Mission," or the cavalry's preparation for an impending emergency. Instead, the top news stories focused on the opening of school, the Marfa-Presidio highway contract letting, a forthcoming football game, and a big illegal liquor haul carried out by the border patrol. The next issue, published on September 24, carried a brief item on page seven noting that the appearance of four army planes at Johnson's Ranch "caused quite a bit of comment locally and news items later appeared in the daily press in the state" and indicating that a "proposed raid from bandits across the Rio Grande had been frustrated by the arrival of these ships. . . . There was good ground for the rumor" that families of ranchmen along the border were abandoning their homes and "fleeing to the towns further inland where protection can be afforded." According to this report, the aviation maneuvers were staged to determine if the Air

Corps could patrol and protect the region from bases in San Antonio and El Paso (*Big Bend Sentinel,* September 24, 1931).

Apparently annoyed by the continuing reports of Mexican bandits threatening the Texas border, Governor Ortiz of Chihuahua retaliated with a verbal bombshell that must have jolted the very foundations of the small West Texas community. The entire matter of the impending bandit raids, the governor claimed, was a hoax perpetuated by "Marfa business men to prevent transfer of American troops in compliance with [United States] war department orders." He disclosed further that the Marfa Chamber of Commerce urged him to appeal to American authorities to retain troops at Marfa "because their presence was beneficial to commerce in towns on the Mexican side of the border." Following his refusal, Ortiz claimed the Marfa businessmen and ranchers "appealed to the American war department to retain troops in that area as a *protection against activities of Mexican bandits.*" In conclusion, the governor reiterated emphatically that no bandits were active in Chihuahua along the Texas border (*San Angelo Standard-Times,* September 28, 1931, italics mine).

Governor Ortiz's accusation apparently went uncontested by the Marfa Chamber of Commerce. No public disclaimer has been found. The governor's claim that the entire matter was a hoax is undoubtedly valid; reports that bandits were threatening the Rio Grande settlements were never verified. A public statement issued by Texas adjutant general William L. Sterling a week prior to Ortiz's disclosure negates further the claims that there was, or ever had been, cause for the emergency. Sterling stated "he had been in communication with personnel along the Mexican border and there was no fear of bandit incursions" (ibid., September 21, 1931). It is noteworthy that the adjutant general's statement is in marked contrast to Governor Sterling's plea to the secretary of war, claiming the removal of Fort D. A. Russell "will open that region to a reign of border banditry."

The "Border Raid Mission" based at Johnson's Ranch, finding no enemy to engage, returned to the San Antonio base on September 19, 1931. The official news release from the Eighth Corps Area headquarters mentioned no emergency. According to a Corps Area spokesman, the aircraft had returned from a maneuver along the border in the vicinity of Marfa "where they had been engaged in advanced airdrome exercises." In addition, the tactical problem of this maneuver was to operate aircraft "from an advanced airdrome where all supplies must be carried from the established bases. Practice reconnaissance flights were made along the Rio Grande" (*San Antonio Express,* September 19, 1931).

The Texas adjutant general's statement disclaiming an emergency was either not communicated to the First Cavalry Regiment headquarters at Fort D. A. Russell or was ignored. In the face of repeated denials, the cavalry embarked on an organized surveillance operation throughout the Big Bend area. John F. Kasper, a member of the Second Signal Company, Second Division, stationed at Fort Sam Houston, participated in the operation. He recalled the entire First Cavalry Regiment was involved in this operation, but "only about a squadron at a time out of Marfa [Fort D. A. Russell]" (interview, September 24, 1979). In addition, various other units were ordered to the Big Bend area, including one of the regiments from the Seventh Cavalry at Fort Bliss. Four cavalry troops constitute a squadron, each with an authorized strength of one hundred men and mounts. While none of the troops operated at full strength, Kasper estimates that at least 250 men and their mounts participated in the search for an enemy that apparently never existed (correspondence, January 13, 1978). (RG 394, including the records of Fort D. A. Russell [Camp Marfa], contains no reference to this operation.)

The patrol units established temporary outposts every twenty miles along the Rio Grande, with a noncommissioned officer in charge. Each morning mounted patrols embarked from these posts, patrolling ten miles in each direction until they met the patrols from the adjacent posts. A shortage of radio operators and equipment in the First and Seventh Cavalry Regiments led to Kasper's Second Signal Company being drawn into the 1931 border patrol operation. Writing of this experience almost a half-century later, Kasper still harbors grim recollections of both the trip to the Big Bend and the lack of accommodations he encountered there. His radio detachment traveled from Marathon to Johnson's Ranch in World War I trucks known as "Class B." "'B' stand[s] for broken springs and hard rubber tires. The 2nd Signal [Company] was a semi-mounted organization and we all wished they had sent our horses with us." Arriving at the ranch, he discovered that the radio unit's quarters consisted of a "covered goat house with dirt floor. . . . We took it because the squad tents had not been furnished the detachment from Fort Bliss."

The signal corpsman also remembered the noticeable shortcomings of the mess accommodations at this temporary river outpost. Kasper noted with some bitterness that Mrs. Johnson's renowned culinary skills, praised by so many Air Corps personnel, apparently were not accorded to other branches of the military. "If Mrs. Johnson ever cooked for us I did not know it," he wrote. Consequently, his brief tenure at the Cooks and Bakers School compounded further Kasper's distaste for his Big Bend assignment. "Besides taking my turn on the radio, I was [also] the cook. Most of it opening cans" (ibid.).

In addition to his unsolicited mess assignment, Kasper's "turn on the radio" did little to enrich his Big Bend sojourn. "In those days," he explained, "the little Signal Corps [transmitter] sets were fixed, like in little suitcases." To generate current for these portable transmitters, the Signal Corps provided an electric generator mounted on a stationary bicycle-like device with an adjustable seat and a "little generator in front that you turned by hand. Just like you would turn an ice cream freezer with two handles. I was the one that generated the juice, sitting on that bicycle seat and just pumpin' that cotton pickin' radio" (interview, September 24, 1979).

The Signal Corps radio unit based at Johnson's Ranch had about a twenty-mile range, "a good day's ride on a cavalry mount," and was capable of reaching the adjacent cavalry outpost upriver. This unit transmitted on a regular daily schedule, and, according to Kasper, "we came on almost like Dr. Pepper, at ten, two, and four o'clock." Also, when the mounted patrols arrived at the station, "the sergeant would get a [radio] report from the patrol leader—from the pickets that came in from both ways—[asking] did they see anything during the patrol that should be reported" (ibid.).

Many of the cavalrymen soon tired of the monotonous patrol routine and, to vent their frustration, began testing their marksmanship on any live target spotted across the Rio Grande. According to Kasper, this was the only border emergency that the cavalry quickly resolved. After each patrol "it . . . [became] the duty of the NCO to count the rounds [of ammunition] of each man on patrol, as the Mexican government was complaining that stray shots were being fired across the border at anything that moved. I would say that there was some truth to this" (correspondence, January 13, 1978).

In his communications assignment, Kasper was privy to official military transmissions. He stated that throughout the period of active patrolling there were no reports of border violations nor was there any evidence of Mexican bandits lurking along the Rio Grande, "none whatsoever. . . . Of course, we were armed with a pistol, but I never drew a side arm" (ibid.). Kasper's on-site observations substantiate further the Chihuahua governor's claim that the so-called emergency stemmed from local economic self-interest. He maintains that the matter was a "conspiracy by the ranchers in the area to bring some income to help the economy of Marathon, Alpine, and Marfa." Elmo Johnson was even more specific. He claimed that "Marfa and Alpine wanted the [Fort D. A. Russell] payroll and El Paso was benefitting from the wholesale business" (interview, June 3, 1966). While Kasper admitted that he "heard nothing but rumor, I do not believe anyone else saw directly any bandits along the river" (correspondence, September 15, 1978).

Although W. D. Smithers was not in the Big Bend at the time of the alleged emergency, his close association with the Mexican people of northern Chihuahua and his knowledge of their day-to-day lifestyle enabled him to view the event with an insight that apparently escaped other observers. "I do know that there was no threats of any raids at that time or later," he wrote. "The Mexicans had such a good thing going, smuggling liquors during Prohibition, that they would have not made raids and spoiled their big profits smuggling" (correspondence, October 9, 1978).

The cavalry patrols continued well into October 1931, Kasper recalled. During this period the riders encountered no Mexican bandits, and despite the causes and outcomes of an action that was undoubtedly embarrassing to many of the participants, the operation did have its lighter moments. One occurred at Johnson's Ranch prior to the Air Corps unit's return to the San Antonio base.

Col. John W. Egan remembered that after returning to the Big Bend airfield from Fort D. A. Russell they met the two *rurales* that had been patrolling the Mexican side of the Rio Grande. Since they also remained overnight at the airfield, an infantryman showed one of the *rurales* how to fire the automatic rifle. Egan explained:

> They took him out away from the houses there . . . and let him lay down on a hump of sand and gave him the automatic rifle and told him to aim at something. He had the strap, the rifle sling, wrapped all around his arm. . . . They coached him and told him to hold it as tight as he could and pull the trigger. Well, the recoil raised him about two feet off the ground with his toes still sticking in the sand. You shoulda seen the look on his face. I'll tell you, talk about wonderment, I'll never forget that. (Interview, February 20, 1979)

There were others who also remembered the event; word about "the rifle that never stops shooting" spread fast up and down the Rio Grande. Weeks after the "Border Raid Mission" departed from the Big Bend airfield, Mexicans trading at the Johnson trading post continued to ask questions about the automatic rifle. Johnson later told Col. Don W. Mayhue that that was probably the only direct benefit resulting from the alleged emergency. After the infantryman let one of the *rurales* shoot the Browning automatic rifle, "word got around on the Mexican side that American soldiers at Johnson's Ranch had a 'rifle that never stops shooting.' Well, that sure put a stop to a lot of that cattle stealing on this side of the river" (interview, February 7, 1978).

Late in October 1931, the so-called emergency was lifted, the temporary river outposts closed, and the various cavalry units returned to their permanent posts. A half-century later a veil of secrecy still shrouds a questionable military operation that deployed a mobile fighting force along one of the nation's remotest boundaries to search for an enemy that apparently never existed. Military records provide no clue to the command decisions, and cavalrymen at troop level profess only casual knowledge of circumstances leading up to the event. When questioned about civilian involvement in the erroneous announcement of the bandit raids, retired United States Army Lt. Col. Samuel G. Smith, then a private at Fort D. A. Russell, explained somewhat reluctantly, "Well, I heard the same thing, that there were stories put out that there was a pending [raid], but . . . that's tied in with what we know at the 'Indian' level in the Army. . . . It may have been that all the officers knew" (interview, January 30, 1980). C. J. Alvarado, postmaster at Redford, Texas, at the time of the purported raids, sustains Colonel Smith's recollections of the matter. "I believe that what they said about the removal of Fort D. A. Russell was right," he wrote, "as I remember quite well hearing the same tale" (correspondence, July 8, 1980). (Contemporary civilian observers in the Big Bend area are either deceased or choose to bear their secrets to eternity. Letters I published in the *Big Bend Sentinel* soliciting information on this event drew no responses.)

International relations between the United States and Mexico during this period produced no causes for disturbances along Chihuahua's northern border. A retrospective examination of local events and conditions preceding the border emergency, however, substantiates further Governor Ortiz's claim.

The close alliance developed between officers of the First Cavalry Regiment at Fort D. A. Russell and their Mexican colleagues in Chihuahua undoubtedly would have provided a communications medium for solving a border emergency had one existed. On December 8, 1929, Col. John S. Fair hosted a luncheon at the First Cavalry Club honoring Gen. Abundino Gómez, Fifth Military District commander, headquartered at Chihuahua City. After explaining that his visit was "to cultivate good friendship between the armies of the two republics," Gómez invited Colonel Fair, the First Cavalry Regiment polo team, and "all the officers and ladies of the 1st Cavalry to visit him" in Chihuahua for an international polo tournament. On the morning of February 22, 1930, Colonel Fair and his official entourage, with War Department authorization, crossed the Rio Grande at Presidio as a sixty-seven-piece Mexican band "gave a beautiful rendition of the Star Spangled Banner, followed by the Mexican National Athem." Joining Colonel Fair on this goodwill visit were "thirty of the representative citizens of the Big Bend, many accompanied by their wives."[1] Eigh-

---

1. *Cavalry Journal* (April 1930): 292–293.

teen months later the guests of General Gómez were patrolling the Chihuahuan border searching for raiding parties that, in all probability, neither side assumed really existed.

Local economic conditions continue to loom large as the motivating force behind the border emergency. By the autumn of 1931, most people living in the Big Bend country had experienced firsthand the sobering reality of financial restriction. The area's single-commodity ranching economy, weakened further by the closing of the Shafter silver mines, appeared near collapse. The price of choice beef cattle had dropped to twenty dollars a head as retail meat prices reached an all-time low. Marfa grocery stores advertised thirty-cents-a-pound round steak, while hamburger meat sold for a nickle (Earney, 1977). And now, faced with the added prospect of losing Fort D. A. Russell, the area's last major payroll, the community approached a crisis.

Organized community campaigns to maintain military posts scheduled for abandonment occur with marked regularity. In the wake of every major conflict from the Civil War to the Vietnam War and its aftermath, the threat of closing obsolete military facilities has touched off a wave of civic emotionalism. This is "natural, human, civilian interest in any community when a military post is taken away," stated Lt. Gen. Samuel L. Myers, a second lieutenant stationed at Fort D. A. Russell in 1931. In addition, local civilians

> may look down their noses at the military while they are there, and they may treat them as second-class citizens . . . which certainly was true in my case in Marfa. But nevertheless, threaten to take them [the soldiers] away, and they are up in arms, organized with every kind of alibi for keeping them there . . . [with a campaign] generated in one of a hell of a big hurry. (Undated interview, 1980)

That the Big Bend community mounted a full-scale campaign to keep Fort D. A. Russell remains a simple expression of economic self-interest. That in the process some proponents expanded their argument beyond valid local concerns projects the event into the broader scope of international involvement. Fear of a return to "a reign of border banditry," the theme articulated by the Texas governor to the secretary of war, remained the campaign's rallying call. The Escobar Syndrome was being resurrected as negotiable currency. Although once a valid claim, the appeal had lost most of its audience by 1931. That the Washington bureaucracy stood firm in the face of mounting political pressure indicates further that the federal agencies as well as "the representative citizens of the Big Bend" understood fully current conditions along the Rio Grande. Isolated thievery did not merit regimental control.

If claims made to the War Department by some Big Bend residents and their political allies strayed somewhat from the facts, the department's responses may also be subjected to scrutiny. General MacArthur's explanation that cavalry staff reductions necessitated the closing of some "small isolated stations" appears valid, as does the secretary of war's explanation to Governor Sterling that changing border conditions rendered the cavalry post obsolete. Highly suspect are General Moseley's responses to Cone Johnson. Although the Air Corps had demonstrated its response capability in that region, the confidence the general placed in "motorized and mechanized units" and the improved highway systems, especially in the Big Bend, is largely visionary (Myers, 1976).

Reference to the economic benefits of closing unnecessary army posts and concentrating the troops at central locations appears both timely and valid; however, such was not the case. The United States Army reorganization and centralization plan then under study was not achieved until after World War II. Neither were the economic benefits realized; only two years after transferring the First Cavalry Regiment and closing Fort D. A. Russell, the Second Battalion, Seventy-seventh Field Artillery, regarrisoned the post.[2]

There was, however, one fundamentally valid reason for closing the post that Secretary of War Hurley had articulated earlier to Governor Sterling: the cavalry post had long since ceased to serve the purpose for which it was established. In addition, maintaining the First Cavalry Regiment at Fort D. A. Russell was an unnecessary and unjustified drain on the defense budget. Lt. Col. Lucas J. Ashcroft, who arrived at the Big Bend post as a private in August 1932, agreed: "It seemed like a high price to pay for a small raid on a few cattle" (interview, January 29, 1980).

Amidst the verbal wrangle over the closing of the cavalry post, the Eighth Corps Area intelligence section continued to look with guarded suspicion toward the Rio Grande. On October 2, 1931, Maj. James A.

---

2. In an attempt to better consolidate his command, the First Cavalry Division commander transferred the Second Battalion, Seventy-seventh Field Artillery from Fort Sill, Oklahoma, to Fort D. A. Russell in 1935, as there were no facilities at Fort Bliss. The former cavalry post was available, conveniently located, and could be reopened at a comparatively small cost. In 1942, however, Fort D. A. Russell again became a cavalry post when the First Squadron, Fifth Cavalry (mechanized) replaced the Seventy-seventh when that unit was finally transferred to Fort Bliss.

Watson, Corps Area executive officer, directed the following confidential memorandum to the Fort D. A. Russell intelligence officer:

It is felt that the movements and plans of the military persons recently deposed from the Mexican cabinet, (named below), will within the near future be of the keenest interest to this section. All sources of information available concerning same are being actively followed.

It is requested that you give the subject the closest scrutiny and report without delay any information considered of value.

General Joaquín Amaro.................. War and Navy.
General Juan Andreu Almazán.......... Communications.
General Lázaro Cárdenas................. Home Affairs.
General Saturnino Cedillo................. Agriculture.

(File S-2, RG 394)

The security concerns that prompted Major Watson's memorandum never materialized and they certainly had no appreciable bearing on the future of Fort D. A. Russell. By the autumn of 1931 the issue was virtually settled; the Big Bend community had challenged the United States War Department and lost. The well-publicized emergency and the ensuing decision to patrol the Rio Grande accomplished little in prolonging the cavalry's tenure in the Big Bend. Forces of technological change were already in motion; creating a local emergency served only to delay the inevitable.

The delay was discouragingly brief. On November 17, 1931, two months and one day following the alleged bandit raids, Maj. Gen. Guy V. Henry, chief of cavalry, directed a memorandum to the commanding officer of the First Cavalry Regiment, Fort D. A. Russell, announcing the mechanization of the regiment.

Under War Department instructions issued this date the 1st Cavalry is, within a few months, to be converted from a horse to a mechanized regiment.

It is with a feeling of sadness that we see this change in our oldest mounted organization with memories of a century's service as such.

However, it is most befitting that the 1st Cavalry should be designated as our first mechanized cavalry regiment. A few years hence its personnel will point to this fact with the same pride as its present personnel cherishes its gallant history since 1832, as the first mounted organization in the Cavalry of our Regular Army.[3]

---

3. *Cavalry Journal* (November–December 1931): 58.

The communication from General Henry conveyed a feeling of finality, allowing no margin for appeal or reconsideration. The Marfa Chamber of Commerce, nevertheless, still regarding maintaining the post as an obtainable goal, dispatched its secretary-president, John F. Robinson, to Washington in January 1932 to wage a "last stand fight to retain some kind of military unit at Fort D. A. Russell." When Robinson's efforts yielded no positive results, the more realistic members of the community began to alter their stand on the matter. The ever-loyal *Big Bend Sentinel*, recognizing this change, shifted its editorial policy and accepted the inevitable: ". . . even the most optimistic of us have conceded that there is little or no chance that the War Department will change its plans" to transfer the First Cavalry Division to Fort Knox, Kentucky (*Big Bend Sentinel*, January 21, 1932).

Even with the First Cavalry Regiment's transfer a foregone conclusion, the specter of Big Bend border troubles remained a nagging concern for the Corps Area's intelligence section. The exchange of correspondence between Maj. L. B. Clapham, assistant chief of staff, G-2, at Fort Sam Houston, and Lt. Grant A. Williams, Fort D. A. Russell intelligence officer, remained brisk throughout this period. The subject: rumors of border disturbances. This prompted Lieutenant Williams to conduct an on-site evaluation in the lower Big Bend region. He included Johnson's Ranch on his itinerary. Lt. Col. Lucas J. Ashcroft, who as a private accompanied Williams, recalled that

> when I arrived [in August 1932], there were still strong barracks rumors of impending raids by Mexican bandits, especially cattle rustling. A short time after arriving I was assigned to drive a Capt. Williams, Hq. Troop, into the Big Bend area. We were accompanied by a guide from the Cav. by the name of Tony Jacques. We remained overnight at a ranch owned by . . . [Elmo] Johnson. . . . It was rumored that Williams was there to talk to the ranch owner about raids from across the border. (Correspondence, December 9, 1979).

Explaining the attention accorded a single rancher living almost 140 miles from the cavalry post, Ashcroft added that "word around the barracks was that Johnson had quite a bit of clout with the army" (ibid.).

If the rumors of impending border disturbances were calculated to alter the course of events, they achieved nothing. Yet in the face of military's unyielding position in the matter, the Texas congressional delegation in Washington continued to pursue what now appeared to be a hopeless cause. In early December 1932, Texas senators Morris Sheppard and Tom Connally made a final appeal to President Herbert

Hoover to reverse the War Department's decision, but to no avail (*Big Bend Sentinel,* December 15, 1932). For the "representative citizens of the Big Bend," the inevitable was finally at hand. The decision was difficult to accept. The transfer of the regiment—546 officers and enlisted men plus families and dependents—meant an immediate reduction in Marfa's 1932 population of 3,909.

Fort D. A. Russell remained active throughout 1932 while the War Department completed plans to transfer the regiment and assign a caretaker unit to the soon-to-be-abandoned post. On the morning of December 14, 1932, Col. William A. ("Flub-Dub") Austin (those who served under him recall that in moments of exasperation he would exclaim, "Oh! flub-dub!"), commanding officer, First Cavalry Regiment, conducted the final mounted formation of the oldest cavalry regiment in the service. The colonel "rode the lines to the strains of 'John Peel' [the regiment's traditional theme song]. . . . Sabers flashed in the bright sunlight as each succeeding platoon approached and passed the colonel" ("First Regiment," 1933: 55–60). The mounted units proceeded to the reviewing stand where they reassembled for a brief address by Colonel Austin. A visiting journalist who witnessed the emotional ritual that followed the colonel's remarks reported:

> Every officer and enlisted man in the regiment dismounted and turned to face his horse. The men stood for a long moment with hands on the polls of their mounts in a silent farewell. Then a lone horse, caparisoned in black, was led in front of the regiment. The horse was Louie, the oldest mount in the historic First Cavalry. (Cox, 1970: 86)

A sergeant led Louie past the final mounted regimental formation in a silent farewell gesture not only to Louie but to the past that he symbolized.

In the afternoon the Fort D. A. Russell Literary Forum held its final meeting. Ironically, the discussions focused on current topics that contained prophetic overtones: "A Resumé of Franklin D. Roosevelt's Life and Achievements," "Adolph Hitler, His Present Status in German Politics," and "The Japanese Attitude Toward the Lytton Report and Its Possible Consequences." Colonel Austin concluded the daylong ceremonies with a final address, stating: "Tomorrow we begin, in earnest, our preparation for the change, which means farewell to the horse" ("First Regiment," 1933: 55–60). This change also meant farewell to Marfa, the fort, and the Big Bend country of Texas. Members of the First Cavalry Regiment would celebrate their last Christmas at Fort D. A. Russell. On January 2, 1933, they embarked for their new post, Fort Knox, Kentucky.

The closing of Fort D. A. Russell and the departure of the First Cavalry Regiment mark an important milestone in Big Bend history. After more than three-quarters of a century of intermittent cavalry surveillance, citizens of that region were no longer protected by a military garrison.[4] The Indian menace had ended in the late 1880s, and present conditions along the Texas-Mexican border no longer warranted a permanent defense force. And should the unlikely need for protection arise, the Air Corps had already demonstrated its emergency response through the facilities of Johnson's Ranch.

The recent developments along the Rio Grande were significant, not for what they achieved, but for the changes they symbolized within the United States military establishment. In that remote stretch of the Big Bend country, two branches of the military met under combat conditions that, in all probability, did not warrant their services. When the matter was resolved, one bid a fond farewell to more than a century of tradition and embarked on a new mission as mechanized cavalry; the other moved toward the threshold of military supremacy.

Neither the professional performances of Air Corps Col. Arthur G. Fisher, cavalry commander Col. John S. Fair, the success or failure of the "Border Raid Mission," nor the cavalry's search for an elusive enemy along the banks of the Rio Grande failed to alter the course of subsequent events. Changes in the structure and performance of the body military transcend the events of Johnson's Ranch and Fort D. A. Russell and were set in motion by forces well removed from the shadows of the Chisos Mountains and the wind-swept plains of the Marfa Flats. A varying complex of events mobilized the two military forces along the international boundary, and they emerge as symbols, not causes, of change.[5]

---

4. Excepting the Civil War period, Fort Davis, established in 1854, remained garrisoned until 1891. The Camp Marfa–Fort D. A. Russell era spans the years 1911 to 1932.

5. For a comprehensive study of changes in military strategy, see Weigley, 1973.

"... we kept a loaded gun in every room"

CHAPTER 9

With the mechanization of the First Cavalry Regiment and its subsequent departure from Fort D. A. Russell, the horse suffered another loss in its enduring battle with technology. Significantly, this occurred in an area where equestrian prowess had long been a symbol of
man's dominance of the region. Spanish cavalry first extended that
country's dominion north of the Rio Grande in the late seventeenth
century, only to be replaced by the mounted Comanche hordes that
forced it to retreat to the inner provinces of New Spain. The United
States Cavalry in turn wrested the vast lands of the trans-Pecos from
the Indians in the late 1880s and returned in 1911 to protect the Anglo settlers from another menace south of the Rio Grande.

The 1920s mark a new era in Big Bend military history. With the
renewal of bandit activities in northern Chihuahua, the United States
War Department dispatched the first military aircraft to the Rio Grande
to bolster border security. This assignment evolved as another proving
ground for combat aircraft, which demonstrated its potential for aerial
surveillance and attack. By the decade of the 1930s, a landmark decision had been made: Big Bend security no longer required the man on
horseback. Johnson's Ranch provided an entrée for a more efficient
and flexible aerial task force that waited less than three hours away.
Mechanization had again exacted its tribute from earthbound horsepower, as the encroaching arm of air power extended its grasp to the
Big Bend country of Texas.

Following the "Border Raid Mission" and the emergency patrols of
the Mexican border, air traffic to Johnson's Ranch showed a marked
increase, more than doubling that of the previous year. Between July
6, 1931, and July 5, 1932, the third full year of operation, reports of
fifty-one landings appear in the field register. One hundred nineteen

personnel based at six different military airfields account for this activity. Flights for the year originated at Brooks Field, Fort Crockett, Dodd Field, Duncan Field, Mather Field, and Randolph Field. The most significant of these flights occurred on October 11, 1931, immediately following the "Border Raid Mission" emergency. On that date two military aircraft from Dodd Field made the first—and probably only—authorized night flight to Johnson's Ranch. The field register contains no details other than that Lt. Thad Foster and Col. Arthur G. Fisher, who led the flight in a Douglas BT-2B, were accompanied by Lt. Ike Ott and Major Gore, a flight surgeon, in a Fokker C-7A trimotor monoplane. Six privates, presumably infantrymen from Fort Sam Houston, are also listed as passengers in the trimotor.

While no official record of this flight exists other than the field register entry, it is highly probable that, in the wake of the recent border emergency, Foster and Fisher were attempting to sharpen their response capability should the Air Corps be called on to secure the United States border along the Rio Grande. The obstacle of the Chisos Mountains alone was sufficient reason to discourage such an operation, but attempting the flight without electronic guidance systems, lighted airways, and landing field boundary and runway lights emphasizes the importance the Air Corps accorded this operation—or experiment.[1]

The Third Attack Group, based at Fort Crockett, dispatched nine flights to Johnson's Ranch during the year, including another Fokker C-7A trimotor flown by Maj. Davenport Johnson, commanding officer of the Galveston Island base. The major identified the flight in the field register as a "Tactical Mission to Tucson" and lists his passengers as two majors, a captin, and two sergeants. Two combat aircraft accompanied the mission. Lieutenant Foster and Colonel Fisher from the Eighth Corps Area headquarters made the flight in a Thomas-Morse O-19E, while a Captain Hines, based at Fort Crockett, flew a Curtiss "Falcon" A-3B attack plane. The swept-wing A-3Bs accounted for seven of the nine flights originating from Fort Crockett during the year.

Following the staffing of Randolph Field on October 25, 1931, that San Antonio–area base began clearing flights to Johnson's Ranch. The field register contains six Randolph Field entries during May and

---

1. Instrument flying instruction was added to the Advanced Flying School curriculum at Kelly Field in 1930. Col. John W. Egan, who graduated at Kelly Field in 1929, recalled that "we were taught to fly by the seat of our pants." To insure this, Egan's instructor placed tape over the bank-and-turn indicator in his training plane (interview, February 7, 1978).

June 1932; all flights were made in Douglas BT-2 basic trainers. An entry for the first civilian aircraft to land at the airfield appears on November 15, 1931. On that date W. D. Mack of Dallas, Texas, landed a Lockheed aircraft, NC 7428 (type unspecified) carrying four passengers at Johnson's Ranch. During the fourteen years the field was operative, only six civilian aircraft entries appear in the field register.

Throughout the year Lieutenant Foster maintained a close scrutiny of the airfield, registering fifteen times at the Big Bend facility. While most of Foster's inspection flights were made in the company of Eighth Corps Area ranking officers (Colonels Gore, Fisher, and Longanecker and Maj. Robert Coker), he also made three flights to Johnson's Ranch accompanied by ranking Mexican Army officers, an action related no doubt to the recent emergency.

Air traffic remained brisk throughout 1932, with an increasing number of flights originating from Randolph Field and Fort Crockett. The aircraft were largely basic trainers, attack, and observation types cleared for cross-country navigational training purposes, which in many cases meant official justification for spending a night or a weekend with the Johnsons.

Following the removal of troops from Fort D. A. Russell, reports of border depredations appeared in the *Big Bend Sentinel* with greater frequency. Whether this was the fulfillment of earlier claims articulated by local civic leaders, a calculated change in the *Sentinel's* reportorial policy, or simply an endemic border condition remains a matter of conjecture. Aaron A. Green, a Big Bend rancher from 1912 to 1944, subscribes to the latter alternative while carefully delineating between simple theft and organized banditry, a key factor in understanding border conditions in the 1930s. According to Green,

> organized banditry was not practiced [much] after 1920. I will say that there always were, and are today in isolated areas near the border, people who have fears [of bandit raids]. . . . Also there were then and still are a good many Mexicans south of the border who will steal cattle or anything else they can get hold of if they have an opportunity. These are just thieves, not raiders. (Correspondence, November 5, 1978)

Indiscriminate use of the terms "bandits" and "raiders" undoubtedly contributed to the anxiety of those living adjacent to the Mexican border. However, the latter-day pattern of border depredations evolved in marked contrast to that of the earlier Pancho Villa era. Between 1914 and the early 1920s, organized bands of renegade revolutionaries attacked border ranches, stores, and trading posts, stealing horses,

A 1932 aerial view of Johnson's Ranch airfield looking southeast toward Mexico. Trading post is at right center; irrigated farm land is at lower right near Rio Grande. Large cottonwood tree where bandits assembled is visible across Rio Grande adjacent to the trading post. Courtesy Smithers Collection.

guns, and ammunition and causing some casualties. The border violators of the early 1930s, however, sought mainly food and clothing. John F. Kasper, former signal corpsman, saw hunger and deprivation as the motivating forces of these incursions. "These were not bandits," he wrote, "but Mexicans, who after several years of drought in the area, were starving and they were stealing cows and goats to eat" (correspondence, January 13, 1978).

Mrs. Katheryn Casner, who, with her husband, Stanley, operated the Chinati ranch a few miles upriver from Presidio in the early 1930s, also supported the hunger theory. "The commissary was burglarized a few times," she recalled, "and sometimes the Mexicans would cross the river at night, steal a cow, butcher it, and carry it back across the river on horseback. But these were just hungry people, not bandit groups." As a dedicated humanitarian who ministered to sick Mexican people along both sides of the Rio Grande (Mrs. Casner practiced medicine without a license), she came to recognize within the Mexican psyche a moral vindication for theft among the Anglo settlements. "There was quite a bit of theft," Mrs. Casner explained, "but within their frame of reference, it was proper and justified. The idea was that we had so much more than they did. If they took something of ours, it wouldn't hurt us" (interview, January 16, 1978).

For whatever reason or justification, border depredations continued. Responding to renewed reports of banditry along the Rio Grande, the Air Corps quickly moved in to fill the void created by the departed cavalry. On February 11, 1933, the Twelfth Observation Squadron at Brooks Field dispatched three Thomas-Morse O-19Es to Johnson's Ranch to scout the area, determine the size of the alleged raiding parties, and establish a display of force to keep them south of the Rio Grande. Lt. Joseph H. Hicks, the squadron's commanding officer, led the three-plane flight with wingmen Lts. L. S. Calloway and G. R. Greer. Three sergeants accompanied the mission as observer-gunners.

The *Big Bend Sentinel* reported that the three airplanes

> scouted the territory of the Rio Grande border and got in a small amount of practice bombing in that section. A rumor was current that the planes had been sent here on special detail to ascertain the number of bandits roaming loose on the border and probably frighten them into the interior of Mexico. (*Big Bend Sentinel*, February 16, 1933)

The *Sentinel* was only partially correct in its published account of the mission. Hicks recalled that while they did no practice bombing, the operation's primary purpose was a show of force. His instructions were

> to go to Johnson's Ranch, set up ground targets there, and use them for target practice. We dove at the targets and used the 30-caliber machine guns, firing forward through the propellors. . . . The idea was to let them [the raiders] know of our presence and I suppose infer [that] Mr. Johnson had this backing. . . . I am sure we stayed two or three days at Johnson's Ranch and flew at least two air missions up and down the river. . . . The flights were low level and our machine guns were loaded and ready to fire. . . . We did land at a very small town up river, possibly 50 or 75 miles from Johnson's Ranch [Presidio]. I recall we just made passes at it,

landed and took off immediately. Purpose, of course, to be observed and to let them know we could land at that location. . . .
All of the O-19s had radio equipment—transmitters and receivers—Morse code was used then, instead of voice. (Correspondence, December 15, 1979)

Although they observed no groups of mounted men along the river,
Hicks believed the mission was successful, as "I do not recall of any
further trouble at that time." He understood that Johnson requested
the air support and either Lieutenant Foster or Colonel Fisher ordered
the mission (ibid.).

While no major raids by Mexican bandits occurred in the 1930s or
thereafter, periodic thefts along the border continued unabated. In
March 1933 a group of mounted men looted the store on the Brite
ranch south of Marfa. They took mainly food and clothing and, according to the Sentinel, "the bandits escaped into Mexico." A few days
later the commissary on the J. L. Sublett farm and ranch, located a
few miles upriver from the Johnson Ranch, "was broken into . . . and
merchandise consisting chiefly of clothing was taken. . . . It is believed that Mexicans from across the river committed the robbery"
(Big Bend Sentinel, March 9, 1933).

While no loss of life resulted from any of these depredations, they
nevertheless awakened memories of an earlier revolutionary era along
the Rio Grande. For those living in isolation, within an area where
law enforcement was largely a matter of personal responsibility, recollections of the past still dominated the present. The Roy Stillwells
ranched on the eastern perimeter of the Chisos Mountains, and Hallie
Stillwell remembered how the specter of fear conditioned her day-to-
day lifestyle.

I rode with the cowboys every day because it was not safe for me to
be alone in the ranchhouse. We kept a loaded gun in every room
and everyone knew how to use them, and when. Law and order in
northern Mexico, "was not," and there was very little protection
on the Texas side. . . . Everyone still had in their minds the raid
on the Jessie Deemer store [in 1916]. Almost everything loose that
can be carried off disappeared like a shadow in the night. (Correspondence, October 10, 1978)[2]

---

2. On May 5, 1916, Mexican bandits attacked the Deemer store at Bo-
quillas. Both Jesse Deemer and Monroe Payne were captured and carried into
Mexico. They were subsequently freed with the aid of the United States
Cavalry.

Even with Air Corps support available at his request, Elmo Johnson shared Hallie Stillwell's fear of the unseen and awaited the unexpected. Situated on the banks of the Rio Grande within wading distance of Mexico, Johnson realized that his trading post, livestock, and commissary increased his vulnerability to attack and, like the Stillwells, he regarded self-protection as the immediate solution to Big Bend law enforcement. To the Air Corps personnel who visited the Johnson Ranch, the ever-present display of firearms became a symbol of life along the border. Maj. Gen. Hugh A. Parker, who first landed at Johnson's Ranch on July 24, 1932, remembered that

> Elmo used to keep a gun, usually a loaded rifle, in every room of that house. I asked him about it one time, and he said, "I might not have time to go in the next room to get one. I want one handy." And he never would clean 'em. And as I recall, he said, "Naw, if you put oil on 'em, you'll get dust on 'em." Which made sense. (Interview, February 8, 1978)

When I asked Parker what Johnson feared, he responded, "Mexicans, I guess, across the river there" (ibid.).

Johnson's poorly maintained ranch house arsenal also caught the attention of Gen. Nathan F. Twining, who, while serving as chairman of the Joint Chiefs of Staff, wrote W. D. Smithers:

> I often think of my first trip down there [as a lieutenant on December 8, 1933]. Mr. Johnson had a rifle standing in each corner of his house. A young armament officer with us didn't think they were too clean, so he asked Mr. Johnson if he could clean his guns for him. Mr. Johnson said, "Yes, clean all but that one in that corner. That's the one I go hunting with." (Smithers Collection)

During the three months following the removal of troops from Fort D. A. Russell, five ranches in the Big Bend area reported raids—or robberies—carried out by small bands of horsemen allegedly from Mexico. With this increasing amount of illicit movement across the international boundary, Johnson recognized the need for a faster method of communication with the Eighth Corps Area headquarters. He felt that, with the nearest telephone located at Castolon sixteen miles away, the only solution to the communications problem was to establish a radio transmitter on the ranch, a suggestion that the Corps Area administration accepted readily. "When he came in and said, 'Let's set up a communications station down there,'" Maj. Gen. William L. Kennedy explained, "we did, so he could report" (interview, January 19, 1978).

The problems encountered in establishing the radio station, however, were in marked contrast to Kennedy's capsule explanation of the decision. Part of the program of acquiring the communications equipment was peculiar to the total operation of the Big Bend airfield. When the Corps Area was unable to provide a transmitter for the proposed radio station, someone located a surplus Signal Corps unit at Maxwell Field, Alabama. Since simple availability was not synonymous with permission to use, procedural steps had to be taken. To clear the way for installing the transmitter at the Big Bend airfield, the necessary volume of official paper was shuffled, and ultimately "the Signal Corps sent sixteen trucks of men and equipment from Maxwell Field to Johnson's Ranch under the pretext of a military maneuver" (Elmo Johnson, interview, June 3, 1966). At the conclusion of the maneuver, the transmitter was transferred to the Air Corps inventory, and the Air Corps then installed the unit and assigned a permanent operator to the Big Bend airfield.

One condition for establishing the transmitter was that Johnson provide the building to house the equipment. He erected it just north of the ranch house about halfway between the runway and the river. According to Maj. Kyle Johnson, a former radio operator assigned to the field, Johnson constructed the building of twelve-by-eighteen-inch adobe bricks and covered the roof with large metal sheets made from flattened fifty-gallon oil drums (correspondence, June 19, 1979).

Three small rooms constituted the communication facility: one for the radio station, weather station, and the operator's office; one for the power equipment; and one for the operator's living quarters. While not the fulfillment of a communication specialist's dream, Major Johnson found the installation and equipment adequate.

The weather station had a minimum of equipment, including a hand operated wet and dry bulb psychrometer, a book of beautiful cloud photos and some charts.

The radio statio had an old style, low frequency 172 kilocycles/ kilohertz transmitter with about 250 watts output. This monster stood six feet tall and five feet wide and with its controls, took most of the space in the office.

The power room had basically a four cylinder gasoline automobile engine coupled with a homebuilt gear to turn a motor generator for charging the battery, and for generating high voltage for the radio. . . . There was no heating or cooling devices, but the weather was comfortable all year. (Ibid.)

No one recalls when the Air Corps established the transmitter, but some of the pilots who made frequent flights to Johnson's Ranch in the

early 1930s agree that service probably began sometime in 1933 following the closing of Fort D. A. Russell. While no transmission log of
the station exists, Major Johnson explained that the "old style, code-
only transmitter was kept pretty busy" dispatching a wide range of
messages and information. He sent dispatches for the International
Boundary Commission relating to the water flow level of the Rio
Grande and reported daily Big Bend weather data to the Eighth Corps
Area headquarters at Fort Sam Houston that were forwarded to the
aviation weather center in San Antonio.

Law enforcement agencies also made frequent use of the Johnson's
Ranch transmitter. According to Major Johnson, members of the
Border Patrol would stop overnight at the ranch, camp in his front
yard, and usually ask him to dispatch their patrol reports via the military radio station. He gathered little information from the border
patrolmen because "they were vague about their business. They don't
tell you anything and they don't ask anything." However, most of
their transmissions were usually "something simple like, 'Captured
two wetbacks near Johnson's ranch,' or 'Captured escaped prisoner.'"
More classified information, especially that relating to international
border conditions, was dispatched in military code, which Johnson
could not interpret.

All messages originating from this Big Bend radio station were carefully monitored by the military intelligence section of the Eighth
Corps Area headquarters at Fort Sam Houston. There the data were
evaluated and submitted for command-level decisions relating to the
surveillance and security of the United States–Mexican border.

While this remote military radio facility opened wider the Air
Corps' "isolated window into the happenings along the border," the
frequent appearance of combat aircraft over the Rio Grande cast a
lengthening shadow of United States military strength across that
international twilight zone where organized banditry once flourished.
Changes in the internal affairs of Mexico plus the Air Corps' ability to
visibly flex its muscles within view of potential violators heralded an
era of change in the Big Bend country of Texas. The three-plane patrol and strafing mission led by Lt. Joseph H. Hicks marked the last
tactical maneuver the Air Corps dispatched to Johnson's Ranch.

Johnson knew from experience that a loaded gun represented the
last resort in self-defense. In case of a confrontation, it made survival
largely a matter of chance. Considering his exposed riverside location,
this was a risk Johnson preferred to avoid. It was upon this premise
that he orchestrated his own personal protection scheme that remains
unique in the annals of border law enforcement. Johnson based his

plan on the Air Corps philosophy that a strong offense was the best defense. Recognizing that a repeated show of force was the best deterrent for the border violators, he turned to his Air Corps friends and was soon employing the finest combat equipment in the United States for his personal convenience and protection.

The recreational opportunities at Johnson's Ranch kept a steady stream of Air Corps officers flying to the Big Bend airfield. Many of these young combat-trained pilots cooperated with Johnson in his self-devised program. The plan was simple: employ United States military aircraft in a blatant show of force in the vicinity of Johnson's Ranch. Lieutenant Hugh A. Parker was one of Johnson's closest friends and made frequent flights to the Big Bend airfield. On these flights Johnson asked Parker, before he landed and after he took off on the return flight, to fly up and down the river in the vicinity of the ranch and trading post "just to let 'em see the airplane" as a visible warning that the military was ever near. Parker believed "it kinda helped him [Johnson] . . . in people coming across and stealing his cattle and goats. Of course, he would tell 'em that we were patrolling the river for him" (interview, February 8, 1978).

Lieutenant Robert S. Macrum was another frequent visitor in the early 1930s, but he, unlike Johnson's other Air Corps friends, used actual force. Macrum would systematically strafe the Rio Grande with machine gun fire as he approached the landing field. Using the forward synchronized .30 caliber machine gun on a Thomas-Morse O-19, "We would simply fly up and down the river," Macrum recalled, "and occasionally pointing the guns down into the river where there was a clear spot, and shooting so that it could be reported back in the hills on the Mexican side." The objective was to display a visible threat "to let the wetbacks in Mexico realize that there was aerial surveillance, and aircraft with machine guns on them were down there" (undated interview, 1979).

Lieutenant John W. Egan was another of Johnson's close friends who often visited the airfield and understood Johnson's concern as well as the solution he developed to the problem. Egan remembered, "There was a bunch of Mexicans living along the river, and if they heard the airplanes flying over they would pass the word through their 'grapevine' that the Air Corps was along the river so the *bandidos* wouldn't come out" (inteview, February 20, 1979).[3]

Some of the more daring Mexican bandits did penetrate Johnson's "Air Corps curtain," and drove some of his cattle across the Rio Grande

---

3. For an explanation of the Mexican "grapevine," see Smithers, 1976: 6.

into Mexico. The loss was usually temporary because Johnson's Mexican friends living along the river aided him in recovering his stolen stock. General Kennedy landed at the ranch one time when Johnson had just returned from Mexico driving a herd of recovered cattle. "Uncle" Tom Miller, a Big Bend border character who was also present, facetiously questioned the authenticity of Johnson's count as well as the true ownership of some of the recovered animals. He asked Kennedy, making sure that Johnson heard him, "You know who is the only man in the whole Southwest who can tell his cattle by the look on their face?" Miller was insinuating that Johnson himself had indulged in a little self-justified international banditry (interview, January 19, 1978).

". . . this merely poured holy water on an excuse to go hunting." —Gen. P. H. Robey

---

Scene 3.
A Brief Interlude: Fun with Elmo and Ada

"... this is a healthy country
if you don't talk too much"

CHAPTER 10

1933 was a rewarding year for Elmo Johnson. Air traffic continued to increase; the sixty landings entered in the field register represent a four-year high for the riverside airfield. Only sixteen flights may be categorized as either administrative or tactical operations. With the exception of Lt. Thad Foster's thirteen inspection flights and Lt. Joseph H. Hicks' three-plane border patrol mission, all other entries represent turnaround flights officially cleared as "cross-country navigation training," "exercises in strange field landings," "improving pilot proficiency," or accumulating flying time to fulfill service requirements. But whatever terminology was employed to give the operations official sanction, the real reason for many Big Bend flights was to visit the Johnsons and enjoy their hospitality.

By 1933 and during the years that followed, dramatic changes occurred in the vehicles the pilots chose for their Big Bend sojourns. (The Air Corps Five Year Development Program ended June 30, 1932.) Newer, faster, and more advanced combat aircraft began appearing at Johnson's Ranch. On March 31, 1933, Lts. John W. Egan and Joseph H. Hicks arrived at the airfield flying two General Aviation O-27s, the Air Corps' first long-range, twin-engine, reconnaissance-light bombardment aircraft. The O-27 had a wingspan of sixty-four feet and weighed 10,545 pounds; the two 650-horsepower Curtiss Conquerer liquid-cooled engines gave this historic aircraft a high speed of 160 miles per hour. Design concepts incorporated later in many World War II medium bombers (especially the Douglas A-20 and A-26 and the North American B-25) originated with the O-27.

Two months later Lieutenant E. H. Underhill arrived from Kelly Field in a Boeing P-12B. (During this period the P pursuit designation applied to fighter-type aircraft.) This was the Air Corps' first-line fighter aircraft in the early 1930s and was the first plane in this category to land at Johnson's Ranch. With a maximum speed of 189 miles per hour, the P-12 series represented the fastest combat aircraft in service at that time. Its engine (five hundred horsepower) and its light weight (2,690 pounds) gave the little single-place biplane (thirty-foot wingspan) outstanding combat maneuverability. Because of its performance capabilities (exceeding three hundred miles per hour in power dives) and trim lines, the Boeing P-12 became the Air Corps' supreme glamor symbol.

The little biplane had one objectionable characteristic. "They would get in a flat spin and all hell couldn't get 'em out," Col. Roy P. Ward explained. Referring to a popular comment that the P-12 was "all engine," he added, "That's what they would say, 'You had an engine in your lap and a feather in your tail.' That's all you had. And a lot of guys got killed in 'em, but that was to be expected." Then, following a moment of silent reflection, he responded with the philosophical outlook of all who survived these formative years of combat aviation: "Of course, it never was going to happen to you" (interview, April 9, 1980).

In addition to its romantic symbolism, the Boeing P-12 series also represented another landmark achievement in the history of combat aviation. According to air force historian Carroll V. Glines, "The P-12 was the hottest fighter in the Air Force [Air Corps] inventory in the [early] 1930s. The lessons learned flying these planes in peacetime were applied with later fighters in actual combat during World War II" (Glines, 1963: 13).[1]

The largest single flight of combat aircraft to land at Johnson's Ranch up to that time arrived on December 8, 1933. The occasion was the Third Attack Group's autumn "field exercise," i.e., a weeklong Big Bend deer hunt as Elmo Johnson's guests. The nineteen personnel arrived in nine Curtiss A-3B attack planes with a Fokker Y1C-14 cargo plane carrying the group's equipment and provisions. W. D. Smithers, who witnessed this unprecedented display of Big Bend air power, reported:

> When they all had landed and taxied up at one side of the field, they composed the largest and the strongest air striking power ever assembled in this part of our border. The nine attack planes were fully armed with guns loaded ready for use. Later that afternoon

---

1. The Boeing P-12 was the last wooden-wing biplane fighter design ordered by the Army Air Corps. Squadron replacements of the faster Boeing P-26, a low-wing, all-metal fighter, were not received until mid 1933.

they demonstrated their skill of aerial gunnery by diving at old oil drums and riddling them with machine gun bullets. (*El Paso Times*, March 23, 1958)

Whether the air crews arrived in a single plane or with an entire squadron, their first destination after landing was the large, cool patio of the Johnson ranch house, where at least forty people could relax and visit in total comfort. The Johnsons kept the field register for all visitors to sign on a table on one side of the patio. The second officer to sign the register that afternoon was Lt. Nathan F. Twining. His good looks, personal charm, and engaging manner attracted him immediately to the Johnsons, and December 8, 1933, marked the beginning of another lifelong Air Corps friendship. Twining, later appointed chairman of the Joint Chiefs of Staff, would return to that remote base many times to enjoy the Johnsons' company.[2]

A half-century later, former Air Corps officers who visited the Big Bend Airways Station still possess the most vivid memories of Elmo Johnson. As a man who lived by the gun, who found comfort in the sound of aircraft machine gun fire, and who survived and prospered in a land of violence and uncertainty, he emerges as a strange paradox in an equally strange land. His reactions to people and events run the gamut of human response from kindness and gentleness to threats of brutal vindictiveness. He was a sensitive and kindly humanitarian and a man tolerant of displays of hostility; a hospitable host, but one who refused the association of those who threatened to despoil the gentle ambience of his riverside domicile. Yet those who knew him best justify his tolerance of violence—either accepted or enacted—as a survival technique in an area where longevity rewarded those who enforced their own codes of justice. Johnson once explained when he first went to the Big Bend that "the few people there carried the law in a holster or slung over their back" (Cox, 1970: 62).

Johnson's physical appearance was incongruous with the violent traditions the frontier region forced him to uphold. Lois Neville Kelley remembered him as

---

2. In an era when military promotions came agonizingly slowly, Twining advanced rapidly through the officer ranks. Captain Twining signed the field register on September 5, 1937, and a year later Major Twining visited the Johnsons for the last time. During World War II General Twining commanded the Thirteenth Air Force in New Caledonia and later the Twentieth Air Force in the Marianas. The capstone of his career came with his appointment as chairman of the Joint Chiefs of Staff, an assignment he held from 1957 to 1960.

a slender little man about 5-8, I guess, on the slight side. . . . [His dominant features were a] tanned face and dark blonde or light brown hair, gray blue eyes, lean face and firm narrow lips. He was genial company, good natured and generous minded. And Ada, his vivacious, plump little wife was all that. She adored Elmo. And he returned the deep attachment. They lived in a frontier setting where they had endured rugged and good times together. They both knew well the world of cities and sophisticated circles, but the ranch and each other was what they treasured. (Correspondence, October 24, 1980)

An admiring nephew remembered his uncle's paradoxical view of woman's role in society. In one respect he was a traditionalist, in another he tolerated Ada Johnson's untimely breaking of the masculine barrier. "Ada would wear pants and boots in a time-frame when it was almost unheard of for a woman," explained David L. Smith. "On the other hand, Elmo would never permit her to do things such as laundry. If they didn't have a Mexican lady to do it, he would do it himself" (correspondence, December 22, 1980). Yet Johnson, the accomplished marksman, who could hit rocks tossed in the air and strike matches at twenty paces with his rifle, took equal pride in his wife's achievements with a gun. "She could shoot might near as good as daddy could," recalled her stepdaughter, Thelma Ashley. "Don't sell 'er short. She went everywhere he went and she hunted at a time when women really didn't do those things" (interview, January 28, 1980).

"Totally hospitable," is the way Gen. William L. Kennedy remembered Elmo Johnson. "He was hospitable in the old tradition. Anyone that came to his place, whether he liked them or not, was a visitor, he was an honored guest. And Elmo didn't like everybody, but was a consummate old school hospitality man" (interview, January 19, 1978). Col. John W. Egan explained that while Johnson had a great sense of humor, he was "a very soft-spoken man. . . . He didn't exaggerate or fool around; he spoke the truth. Not at all loud or boastful, in a way . . . he was very modest." Although Johnson exhibited a casual and informal bearing, Egan categorized him as "one of the most determined men I have ever met. . . . One would have to be, to establish that ranch in such a remote area and survive. . . . You knew he had a very firm character and he was ready at any instant to defend his interest" (interview, February 7, 1978).

Circumstances frequently dictated that Johnson defend his interests, and in most instances his success depended on his skill with firearms, which he turned to as a last resort. He recalled one instance when he evicted a man, reported wanted by the law, from his property. This person eventually made his way to Alpine where, in an apparent attempt to place Johnson in jeopardy with the law, he told the sheriff that "he wanted to file charges on me for shooting at him . . . [claiming] I had shot at him three times." One of Johnson's acquaintances who overheard the complaint denounced the charge as false, citing Johnson's marksmanship as evidence. He later reported the event to Johnson, saying, "Elmo, I told 'em it was a damn lie! You wouldn't have shot but once, anyway." Although Johnson claimed he "never fired a shot and missed," he maintained he was never driven by desperate instincts. "I never shot to kill [a person], though. I never had to and I wouldn't have wanted to" (Cox, 1970: 62).

Quiet, determined, and totally self-contained, Johnson was regarded by his many visitors as the personification of the regional prescription for survival: "This is a healthy country if you don't talk too much." Col. Elmer P. Rose remembered that on one flight to the ranch he learned that Johnson's prize yearling stud horse had been killed while being pastured on an island in the Rio Grande. Johnson explained that someone had castrated the young animal, "leaving him to die in unspeakable agony in the hot sun." When asked what he planned to do about the matter, Johnson replied, "Oh, nothing now. We know who did it. Some day he'll get a bit careless. We'll get him and he'll know how my beautiful stud died." "Yep," Rose concluded, "Elmo was a man to ride the river with" (correspondence, April 30, 1979).

Although Johnson was understandably revolted by the emasculation of his prized yearling stud, there is no evidence that he ever avenged the atrocity. His threat was issued, no doubt, in the immediate wake of the senseless killing. To Johnson, that inhumane act violated his belief that both humans and animals were entitled to an existence in that uncrowded land and should be free of all encumbrances sometimes imposed by less sensitive individuals. And while Johnson was forced to adjust to the mores of a hostile environment, those who knew him best maintain he remained the same hospitable, soft-spoken, gentle person who first descended on the valley of the Rio Bravo in the autumn of 1927.

Johnson's circuitous trek to the Big Bend began in 1889 in Fannin County, a North Texas agricultural community where he was born. The family lived briefly in the Indian Territory (Oklahoma), where Johnson's father and uncle were engaged in a joint business enterprise. When he was fourteen years old the family moved to Dallas, where Johnson later worked briefly for Dunn and Bradstreet. He married Annie Lee Drake while living in Dallas, and their only child, Thelma, was born there. Shortly thereafter their home burned and, instead of rebuilding, the family decided to move west and establish residence on two sections of land near Sonora, Texas. This was inherited property Johnson's grandfather had received for service in the Republic of Texas army.

In the meantime, Johnson's wife died of diabetes. He subsequently met and married Ada Keaton Morris, a young widow whose father, George W. Morris, owned property near Johnson's Sonora ranch. Johnson ranched that property until 1918, when he sold his holdings and reported to San Antonio for induction into the army. The Armistice, however, interrupted Johnson's military plans, and he elected to remain in San Antonio and enter the real estate business. Using the money he received from the sale of his ranch, he speculated heavily in residential construction. An economic panic in the early 1920s, however, spelled doom for the West Texas entrepreneur. He later told a nephew that when the venture failed he sat on the bank of the San Antonio River and burned a small fortune in worthless second lien notes.

Johnson, however, emerged a survivor. Following the collapse of his real estate enterprise, he decided to forsake the urban sprawl for the vast unoccupied spaces of the Southwest Borderlands. He announced to his wife that "they were going to do a little pioneering. 'I wanted to see the Big Bend,' he said. 'I liked the thought of big, open country'" (Cox, 1970: 63). However, Johnson's immediate problem was capital, and with the legal guidance of his brother Alvis Johnson he salvaged a portion of his original real estate investment. Using this capital interest, he entered into a partnership with Alvis and his father, Robert Johnson, and the three of them embarked on the Big Bend ranching venture.

The site Johnson selected "to do a little pioneering" was a tract of land originally owned by two young men from Kentucky, G. N. Graddy and W. B. Williams. Although the two Kentuckians acquired the property in 1924 and established a successful border trading post, they failed in their effort to develop a tobacco farm on the adjacent Rio Grande flood plain. Their Big Bend sojourn was brief; subsequent to their departure the Johnsons acquired the property through a San Antonio land agent.

Photographer-journalist W. D. Smithers was one of the first people to greet the newcomers to the Big Bend. Smithers had been making regular trips to the Big Bend six years before the Johnsons arrived. "On my 1928 trip [probably a year earlier] when I reached Castolon," he recalled, "the Johnsons were there and we all went down to their new home." This acquaintance led to a lasting friendship and Johnson's ranch became Smithers' Big Bend headquarters for a number of years (correspondence, August 27, 1975; July 5, 1979). (Smithers lived with the Johnsons from August 1928 to May 1930 and for all of 1932.)

Soon after Johnson acquired the property and opened the trading post, it appeared he had launched another financial disaster. Sheriff W. N. Gourley placed a public notice in the May 3, 1929, Alpine *Avalanche* announcing the public sale of the Johnson property, which totaled some one thousand acres. The sheriff acted pursuant to a judgment issued in DeWitt County against Robert Johnson, then listed as half-owner of the Big Bend property.[3] The legal maneuvers remain unclear, but according to David L. Smith, the new owners apparently defaulted on some of their financial obligations relative to acquiring the property and this led to the suit. "There was a foreclosure," Smith explained, "and Mr. Morris, Elmo's father-in-law, apparently bought the ranch at the foreclosure sale and gave it back to Elmo and Ada, thus cutting out his brother Alvis and his father, Robert, which resulted in a long-standing breach in the family" (correspondence, December 18, 1978). Coincidentally, Morris concluded the transaction on July 5, 1929, one day preceding the dedication of the landing field.

After establishing his Big Bend residence, Johnson diversified his interests between the ranch, the farm, and the trading post. While all three contributed to the quality of life the Johnsons enjoyed, only the ranch and the trading post yielded consistent profits. The arid ranch land on which he ran a small herd of cattle limited his profit potential, and while the irrigated farm yielded bumper crops of cotton, melons, grain, and farm produce, the great distances to market made this operation economically unfeasible.

In addition to being a consistent moneymaker, the trading post became a major link in a chain of border trading institutions situated on the Texas side of the Rio Grande between Boquillas and Candelaria. Economic necessity determined their location. With northern Mexico virtually an isolated hinterland, people living in that region were forced to seek markets along the international border. By maintaining a large inventory of merchandise, these border trading posts lured a steady stream of customers—purchasers, sellers, and traders—from the interior of Mexico, thereby insuring the owners a substantial volume of trade.

Most of the transactions were barter and trade; Mexican furs constituted the major trade items at Johnson's trading post. Johnson purchased the heavy timber wolf pelts for from $1.00 to $1.50 each, and he sold them on the American market at approximately three times his cost. He recalled that the fur trade remained brisk throughout most

---

3. County Clerk's Records, Brewster County, Texas, Vol. 67:602.

of the 1930s, with annual profits varying from $3,000 to $10,000. Johnson also pointed out that even with that volume of trade little money changed hands, because the furs were accepted as a medium of exchange. This was doubly beneficial to Johnson. In addition to being assured of his customers' retail trade, this eliminated the need for maintaining large cash reserves at the trading post. With the nearest banks more than one hundred miles away, that was an important consideration.

Johnson discovered early on that while all of the Mexicans' profits from fur sales were usually exchanged for merchandise, they preferred to see the actual money involved in the transaction. To please his customers, Johnson purchased "a water bucket full of dobie dollars" (Mexican silver pesos) and kept it on the counter during a fur trade. As he purchased each pelt, Johnson handed the seller the cash representing the transaction. With the sale completed, the Mexican traders began purchasing their supplies, and "the money started coming back to my side of the counter," Johnson explained. "My store was about the only place these people traded, and nobody used that money except me and my customers. It just kept coming back. The same money circulated for years" (interview, June 3, 1966).

While Johnson traded mostly for furs, poultry, and livestock, his customers could choose from a fairly wide selection of merchandise. He sold ammunition in various calibers, but no guns. Food items were limited to those not produced locally: sugar, coffee, rice, and *canela* (cinnamon) in bark form, with all items stocked in bulk and usually dispensed in small amounts. Cigarette tobacco was a priority item with the men, and *Lobo Negro* (Black Wolf) was the popular favorite. According to Smithers, "Some of that rolled up in a piece of corn shuck, made a smoke that few Americans could enjoy. It was *muy fuente* [very hot]" (Smithers, 1961: 48).

Johnson's customers came from as far as fifty or sixty miles into the interior of Mexico and arrived on foot, on burro, on horseback, and sometimes in wagons. Most trading occurred in the late afternoon. One Air Corps pilot, Col. Charles Deerwester, observed, "It was an odd store and when you would walk in you'd wonder where the customers came from, but about four o'clock in the afternoon they would always seem to show up. You'd never see them cross the river, because they apparently crossed above or below his place, whether by design or accident, I don't know" (interview, February 11, 1971). According to Johnson it was partially by design, because during his Big Bend tenure periodic changes in immigration policy influenced when and where the Mexicans traded.

When Johnson first opened the trading post, "the Rio Grande was

Busy day at Johnson's trading post. Mexican trappers deliver furs on burros. Elmo Johnson stands beside the truck checking the furs for delivery to a buyer in Alpine. Courtesy Smithers Collection.

just a river. The Mexicans came and went as they pleased," he reminisced. In 1934, however, the United States enacted more stringent immigration laws. This statute designated only one day each week as Port Day, during which the Mexicans could cross the river to trade provided they carried a passing card. This regulation gave Johnson a trade advantage over his competitors at Terlingua and Castolon. With his location adjacent to the river, "they could slip over undetected and trade without fear of being caught" by the *fiscales*, Mexican government officials (interview, June 3, 1966).

Although the location of the trading post proved to be a distinct economic asset, Johnson's unfamiliarity with the cultural traditions of his customers became a temporary liability. Experience and declining sales enabled him to grasp the two keys to success in the world of Mexican border commerce: first, learn to speak the language, and second, do not give the Mexicans credit. Achieving the first enabled him to comprehend the logic of the second.

When Johnson first moved to the Big Bend he had little experience working with the Mexican people. By extending credit to the Mexican fur traders, he unknowingly lost not only the credited merchandise but some of his customers and the source of raw furs as well. To avoid settling their overdue accounts with Johnson, they began bypassing his trading post and, instead, took their business to Wayne Cartledge's "cash-and-carry" store at Castolon. Johnson learned quickly that credit bred enemies, not friends. It would be the latter that greatly enriched the Johnsons' new lifestyle by helping spread a spirit of mutual trust and understanding along both sides of the Rio Bravo.

Fluency in the Spanish language enabled the Johnsons to gain a new awareness of the Mexican people and their customs. They achieved this largely through Mrs. Johnson's initiative. She engaged a Mexican girl, who had worked for an Anglo family at Terlingua and spoke English fluently, to help her with the household chores. As they worked together, they conversed only in Spanish.

As the Johnsons gained a better understanding of the Spanish language, they in turn shared with their customers and neighbors some of the more practical skills of Anglo society. Mrs. Johnson taught the Mexican girls to cook, sew, can fruits and vegetables from the garden, and to read and write. "The cookbook and the Sears [Roebuck] catalog were their primers," Johnson recalled (interview, June 3, 1966).

Smithers accords Mrs. Johnson high praise for her effort to enrich the lives of the Mexican women and children on both sides of the Rio Grande. For the Johnsons' first Christmas in the Big Bend, she

> invited all the children to her home to see for the first time an American Santa Claus and a Christmas tree. She, with the help of her cook, Maria, worked every afternoon for weeks preparing for the event. She had toys and gifts for every child and parent—the first they had ever received. Each child received enough candy, nuts, home-made cookies, and fruit to last them several days. She became famous for her Christmas Eve party. (Smithers, 1961: 48).

Johnson soon recognized that the Mexican children deserved more education than Mrs. Johnson's "cookbook college" provided. When he discovered that thirty-five Mexican children living near his ranch were enumerated on the Brewster County school census rolls but were receiving no schooling, he embarked on a one-man crusade to bring free public school education to the lower Big Bend. Johnson appealed first to the local school board at Terlingua with no results, and received a similar response from a county judge in Alpine. As a last resort he traveled to Austin to see State Superintendent of Schools Dr. L. A. Woods. There he received a sympathetic hearing. With an air of justified smugness, Johnson recalled, "We got the school. The teacher lived with us and I furnished the school house" (interview, June 3, 1966).

As the years passed the Johnsons gained a very special place in the social, cultural, and economic life of the lower Big Bend, especially with the Mexican people. The trading post offered both a market for what they produced as well as a shopping center for their personal and household needs. Johnson's fifteen-acre irrigated farm also became a source of fresh fruits and vegetables for all people living along the river. Each Saturday afternoon the Johnsons and their neighbors assembled in the Johnsons' garden where they all worked together and shared what they grew. The Johnsons soon learned that kindness, honesty, and respect were the human qualities to which the Mexican people responded most readily. This was readily visible to all who landed at the riverside airfield. Gen. P. H. Robey visited the Johnsons frequently while stationed at Fort Crockett. He remembered a young Mexican boy who came across the river to sell some farm produce at the trading post. After Mrs. Johnson paid him in cash "she also gave him some flour [and] sugar . . . patted him on the back, and he went back across the river. I sorta gathered then that the river merely separated good neighbors" (correspondence, November 14, 1978).

As the bond of friendship between the Johnsons and their neighbors across the Rio Grande grew stronger, the Mexicans began to adjust their needs in accordance with the new social and cultural influences the Anglo newcomers brought to the region. The Johnsons' contacts with the Mexican people ran the gamut from the very practical to the highly personal, each giving an added dimension to a primitive lifestyle heretofore remote from the outside world.

Many came to the trading post solely for help and advice. Gen. Hugh A. Parker remembered one regular visitor only as El Bandito, a Mexican farmer and goat herder who lived two miles downriver from the Johnsons. He would always appear at the trading post on horseback, clad in overalls and white chambray workshirt, wearing a red bandana tied around his neck and the traditional straw sombrero, and always carrying a leather riding quirt. He rode on a miserable-looking hand-carved wooden saddle on which was tied a hand-plaited leather lariat. El Bandito carried a .30-30 caliber rifle that always needed repairs. For a people who subsisted largely on wild game, the rifle was an essential item. Parker explained that

the main spring on the hammer mechanism was broke and Elmo took an old shoe . . . , removed the arch support, filed it down, and put it in there. Of course, it was about four times stronger than the real spring. When that hammer went forward, boy, there was no doubt about it, it was gonna pop that cap. He was always comin' over with some problem for Elmo, and he'd patch it up for him and do the best he could by him. (Interview, April 13, 1978)

As the Mexicans' confidence in Johnson's sincerity increased, their demands for his services grew accordingly. In the absence of a convenient court system, Johnson eventually found himself offering the Big Bend's only free legal aid. He was called on to arbitrate all manner of personal differences. Sometimes it was the ownership of a watermelon crop, a plot of ground, or a goat or a burro; sometimes it was a domestic upheaval. Whatever the dispute or whoever the litigants, they usually accepted Johnson's verdict as final.

Whether in the presence of the Mexican children, whom she adored, or visiting Air Corps officers, whom she held in high esteem, Ada Johnson projected a positive and enduring impression. Forty-three years after his first visit, Gen. William L. Kennedy's recollections of Ada Johnson remained vivid:

That woman was quite something. She was a most unusual woman in that she had the appearance—in fact, the old Texas twang—of a backwoods woman. But at the same time she was very astute about the history and archaeology of that area. She didn't go up and dig in those caves, but the Mexicans, who loved her, would bring remnants of artifacts in to her, and she could reconstruct that material with all the skill and knowledge of a trained archaeologist. Yes, that woman was quite something. (Interview, January 19, 1978)

A half-century after the airfield was established, scores of retired Air Force officers who visited Johnson's Ranch still harbor lasting impressions of that experience, all pleasant and nostalgic. Second only to their recollections of the character and charm of the Johnsons is Mrs. Johnson's cooking. While most pilots arrived with gift items of food to compensate the Johnsons for their hospitality, Col. Elmer P. Rose felt that many times the great number of uninvited and unexpected military visitors to the airfield was an imposition on the Johnsons. Rose claimed that "many of them went out there without invitations and expected Elmo to take care of them." Then, reflecting on that experience and Johnson's ongoing love affair with the Air Corps, Rose added that Johnson "would not let them know about it" (interview, January 28, 1980).

Although the airfield had radio communications after 1933, the transmissions were confined largely to administrative, weather, and

G-2 intelligence reports, with seldom any information received on incoming air traffic. This presented a very practical problem as Mrs. Johnson never knew who or how many were coming for dinner and who would remain overnight. She soon devised a tentative solution, which Smithers reported as follows:

> The Johnsons had no way of knowing when none or ten planes would [arrive] . . . as there was no telephone, but about half an hour before the plane landed the motor would be heard and her kitchen helper, Andrea, would start things going in the kitchen, and from the door [of the kitchen] she would watch the planes come in [to land] to know how many hungry men to prepare for. (*El Paso Times*, March 23, 1958)

Mrs. Johnson and her cook prepared the food in an adjacent adobe building and transferred it to the main house. There, on a large table that would accommodate about a dozen people, she served the food family style. Almost four decades after being assigned to Johnson's Ranch, Maj. Kyle Johnson still remembered his first meal with the family. It consisted of "goat meat, tacos and beans, beautiful tomatoes, lettuce, and green onions from the garden. . . . Mr. Johnson liked to hunt and we had deer meat, wild pheasant and quail to supplement our regular goat meat. I liked to fish and sometimes I'd bring catfish from the river and Lipa would fix it with ground corn and fresh green onions." Johnson added parenthetically, "We ate well" (correspondence, June 19, 1979). Gen. P. H. Robey supports Major Johnson's high appraisal of the Johnson cuisine. "A real five star restaurant" (correspondence, January 24, 1978).

The relationship that developed between the Johnsons and the Air Corps also encompassed matters totally unrelated to the operation of the airfield. And while the Johnsons may or may not have possessed administrative clout, when they spoke the Air Corps usually listened. A case in point is the military career of Robert D. Johnson, Elmo and Ada Johnson's nephew.

During one of Col. Arthur G. Fisher's frequent trips to the Airways Station, Ada Johnson questioned the air officer about Robert's opportunities for advancement in the Air Corps. At that time he was a private attached to the Fortieth Bombardment Squadron at Kelly Field. Following the colonel's negative response, Mrs. Johnson expressed her personal concern for what she hoped would be a promising military career for her nephew.

According to Thelma Ashley, Elmo Johnson's daughter, the colonel agreed to look into the matter when he had the opportunity. Shortly thereafter Private Johnson was surprised to receive a telephone call from Colonel Fisher inviting him to join a group of Air Corps officers for a weekend flight to Johnson's Ranch. In an era when fraternization between officers and enlisted men was not the accepted mode even in the Air Corps, young Johnson's astonishment is understandable. He recalls they made the flight in a Douglas O-38, left Kelly Field on a Friday afternoon, and returned the following Sunday afternoon.

Johnson's military career was not mentioned during the weekend visit; however, shortly thereafter he received a telephone call advising him of forthcoming examinations for admission to the Radio Operators and Mechanics School at Chanute Field, Illinois. Although his acceptance to the Chanute Field training center was never discussed with his family, Robert Johnson, now a retired lieutenant colonel, still believes his aunt's intercession with Colonel Fisher provided the initial step toward a successful air force career (interviews, November 1980 and undated). "Beyond that point, Robert did it on his own," Thelma Ashley explained. "She [Ada] just opened the door for him" (interview, January 2, 1982).[4]

During their years in the Big Bend the Johnsons enjoyed the best of two worlds. The remote southwestern frontier offered the companionship of the peoples of two nations as well as independence, abundance, and freedom from the constraints and encumbrances of modern society. And while relatively few people challenged the unimproved roads that led to their Big Bend ranch, the Johnsons were seldom alone. The outside world remained always near. The radio station linked the airfield to the Eighth Corps Area command center at Fort Sam Houston, and the runway that extended from their front yard served the finest military aircraft of that day. These facilities offered the region immediate protection and attracted a growing host of military friends and associates. In addition, Elmo Johnson's accreditation with the United States Army Air Corps gave him access to the Air Corps planes, a privilege that he frequently expanded to include his personal needs and conveniences. "I was sitting pretty high," he once explained. "I always had overnight guests, and as time went on I got free use of military planes and good trade" (Cox, 1970: 63).

---

4. Following graduation from Chanute Field, Robert D. Johnson served as instructor in instrument flight procedure and celestial navigation. Shortly after his promotion to warrant officer in May 1942, he received a field promotion to second lieutenant. He entered flight training at Kelly Field, flew combat missions as a B-24 pilot in the Italian campaign, and later became a squadron commander. Following thirty years of service, he retired as a lieutenant colonel.

Advantages sometimes breed disadvantages. Johnson's unsolicited role as reluctant potentate of his river empire frequently placed him at odds with his urban-based colleagues at Marfa and Alpine. Col. Charles W. Deerwester, a frequent visitor to Johnson's Ranch, recognized this growing animosity. He recalled that people from Alpine and Marfa "were a little bit jealous of Elmo and his wife and their free life down in the valley there. As I recall dimly back in those days, Elmo had some problems up there" (interview, February 11, 1971). Jealousy may have been a factor, but part of Johnson's problems stemmed from his remote location at the southern tip of Brewster County. Traversing 150 miles of poorly maintained roads to the border trading post prompted some Alpine wholesalers to reject Johnson's orders. Understandably, he interpreted this as an unfriendly act that merited appropriate retaliation. "Alpine boycotted me and I boycotted them," he recalled years later. "I trucked all my supplies from San Antonio" (interview, June 3, 1966).

Johnson's relationship with the Alpine civic leadership fared no better. In the late 1930s, Herbert Meyer, regional director of the National Park Service, arrived at Johnson's Ranch by automobile accompanied by several Mexican government officials and a representative of the Alpine Chamber of Commerce. Meyer wanted to use an Air Corps airplane for an aerial inspection of the proposed Big Bend National Park site and filed an official request via the military radio facility at the ranch. Meyer's request was summarily rejected, as was a subsequent appeal issued in the name of the Alpine Chamber of Commerce. As Meyer's party was preparing to depart, Johnson suggested that he might be able to provide them with an Air Corps plane. According to Johnson, an indignant Meyer exclaimed, "If I can't get a plane, how the hell can you?"

After suggesting that Meyer delay his departure thirty minutes, Johnson sent a message to the chief of the Army Air Corps in Washington requesting permission to fly Department of the Interior and Mexican government officials over the park area. Johnson received a prompt answer. It read: "Request for plane from Dodd Field granted." Johnson, recalling the incident with a smug grin, explained, "That didn't rate me with the Alpine Chamber of Commerce a damn bit" (ibid.).

Johnson once admitted that he "wouldn't have won any popularity contests at Marfa either." Following the closing of Fort D. A. Russell, the War Department ordered the demolition of certain temporary facilities designed to serve the horse cavalry. Johnson felt that, considering his Air Corps affiliation, it was ironic that he receive the army contract for this work, which in addition allowed him to salvage materials for his personal use at the ranch. But what Johnson salvaged was not always what Johnson got. Much of the lumber he recovered during the day "would be hauled away during the night by Marfa area ranchers. They thought they were hurting me. I didn't care and neither did the Army. They just wanted that damn thing tore down before the politicians could stir up enough stink to get it stopped." Johnson, however, salvaged enough material from the former cavalry post to erect a ten-room barracks building at the airfield to accommodate transient Air Corps pilots (ibid.).

Along the Rio Grande Johnson functioned in a world apart from the urban villages that dotted the Southern Pacific railroad some 150 miles to the north. His day-to-day contacts were with people who had no concept of a competitive capitalistic lifestyle, whose world of poverty was limited to material possessions, and who looked upon the Johnsons as neighbors and mutual survivors in a harsh, unyielding land. The land and the river and the trading post brought Johnson and his neighbors together, and each gave to the other a richness of experience that neither could have found alone. One was Mexican, the other was Anglo, but by the mid 1930s the river merely separated good neighbors. To them, the Escobar Syndrome was only a bad memory.

"... most pilots were pretty well disciplined in the military"

CHAPTER 11

Approximately 350 air miles separated the San Antonio–area Air Corps bases and Johnson's Ranch. And whatever technical skills a young Air Corps officer might wish to develop—cross-country navigation, strange field landings, or increasing his flying time—the approximately three-and-one-half-hour flight over some of the Southwest's most varied terrain offered ample opportunity to fulfill his training needs.

The air route the pilots followed to the Big Bend airfield presented diverse navigational problems. As all flights to Johnson's Ranch were cleared for daylight contact operation (flying with visual reference to the ground), the San Antonio–Dryden segment posed little challenge to the pilots' navigational skills. (On the return flight, the Dryden–San Antonio segment was sometimes flown at night.) They simply followed the Southern Pacific railroad tracks west and checked their progress with the passage of Uvalde, Brackettville (Fort Clark), and Dryden. According to Gen. William L. Kennedy, following railroads was an accepted practice in the 1930s. "We did it all the time and that Southern Pacific is a pretty straight railroad" (interview, January 19, 1978).

The Air Corps maintained an auxiliary field just west of Dryden, staffed by a sergeant who helped refuel transient military aircraft. With the short cruising range of the airplanes flying to Johnson's Ranch in the 1930s, a fuel stop at Dryden, usually on the return flight, was mandatory. The army later added a hangar with limited overnight facilities for Air Corps personnel who became stranded because of darkness, mechanical trouble, or weather.

For the young Air Corps pilots seeking cross-country navigational experience, the Dryden–Johnson's Ranch segment of the flight was indeed a challenging course. Four decades later Maj. Kyle Johnson, the former radio operator stationed at the airfield, recalled that "it would sharpen the pilot's navigational skill just to find it [Johnson's Ranch] in the first place, being a small speck hidden behind a mountain, and further sharpen him up to land in the sand with a crosswind, while watching out for those greasewood stumps" (correspondence, June 19, 1979). Col. John W. Egan, a former navigation specialist and close friend of the Johnsons, defined the problems of locating the Big Bend airfield in terms of dead reckoning, compass headings, and geographic check points.

> From Dryden you had to navigate by your compass and dead reckoning down to Johnson's. That was 150 miles, more or less. . . . We had maps that showed the mountains. You would navigate until you saw the Little Dove Mountains in that grey plain, which was fifty miles southwest of Dryden. And then you would pick up the Santiago Mountains and you crossed that ridge at a certain angle. Then you would be on the Deadhorse Mountains, then you would be over the Tornillo Flats, and about that time you would see the river [Rio Grande] on your left by Boquillas. It would be off a little ways. Then you would see the Chisos [Mountains] coming up on your right. You would stay east and south of the Chisos, and then you would be able to see the river [Rio Grande]. You would follow it on around until you could see the landing field at Johnson's coming up on your right. (Interview, February 7, 1978)

Flying time to Johnson's Ranch varied according to the aircraft's cruising speed, the cruising altitude, and the prevailing wind—headwind, tailwind, or crosswind. The Douglas BT-2 and the Thomas-Morse O-19, for example, had an indicated cruising speed of beween ninety and one hundred miles per hour. The prevailing winds on westbound flights frequently reduced the ground speed to around seventy-five miles per hour. The time lost on the flight out, however, was regained on the return trip. Usually a pilot with a passenger and/or observer could log about seven hours of flying time on a flight to Johnson's Ranch. If two pilots made the flight together, they divided the flying time equally.

Comfort and safety were key factors in selecting a cruising altitude. Egan recalled that they usually flew out at around nine thousand feet, but if the weather was hot they would climb to around thirteen thousand feet. "Not much higher than that. The airplane would not perform real well up there." The higher the altitude the greater the margin of safety if the pilot had to make a forced landing. Since most aircraft flying to Johnson's Ranch carried maximum load (fuel, passengers, and baggage), a fast landing on rough terrain was something to be avoided. "If you had a forced landing [loaded], your plane would sit down kinda fast and hard and you wanted as much altitude as you could get in order to pick out a soft spot," Egan explained. "They were few and far between once you left Dryden" (ibid.).

Forced landings were a way of Air Corps life—and sometimes death. With the high incidence of engine failures in the early 1930s, the training command developed a procedure—exercises in strange field landings—to better equip the pilots for this eventuality. Col. Roy P. Ward, who graduated from the Air Corps flying school in November 1931, explained that after the students became proficient in all primary maneuvers

> they would take us out over the open fields southeast of Brooks Field and give us an exercise called strange field landings. We would go out in formation [the instructor and four student pilots] and the instructor would break off and pick out a field [a pasture, a hay meadow, or any open unplowed field] to land in. . . . Most of the instructors knew what fields were safe. . . . In those days the farmers didn't give a damn. You could land and take off and nobody would pay you any mind. So then the instructor would land in a field to demonstrate, taxi back, and take off again.
>
> We would be circling in formation over the field. He would come back and join us. So then he would tag along on the end [of the formation], and the leader [a cadet] would then pick out a field somewhere else and he would make a landing. If his landing turned out to be OK—or it looked like it from the air—all the others would follow him in. Then we would all take off and go find another field. (Interview, April 9, 1980)

This exercise, repeated frequently, was always conducted under the supervision of an instructor. Ward believed this training was essential as "in our day and time it was almost an everyday occurrence that somebody would have a forced landing. We had a tremendous incidence of engine failure" (ibid.).

Lt. Robert S. Macrum's training exercises in strange field landings served him well. Just east of Dryden, Texas, on a flight to Johnson's Ranch in a three-plane formation of Curtiss O-40s, "I got a frozen bearing and had to drop out of the formation for a forced landing," he remembers. "Fortunately, right there was the Southern Pacific railroad underneath us and I found a spot along the tracks where I put in." After seeing that Macrum and his passenger had survived the landing, the two remaining aircraft continued on to the auxiliary field at Dryden to report the incident and summon help.

The Southern Pacific railroad remained the Air Corps pilot's best friend. After alighting from his crippled airplane, Macrum discovered a railroad telephone box nearby and reported his location to the company operator, who, in turn, relayed his message to the Dryden facility. The staff at the auxiliary field responded to the emergency posthaste. Considering the chain of events, the young lieutenant was far from delighted with the Air Corps' expeditious response to his call.

> They were too damned efficient at Dryden, and got a truck and a crew out to me before evening. I say unfortunately because some nearby rancher had appeared and invited me to a box supper that evening. Oh, well, that's the way it goes. I had visions of having a cute young chick for a supper partner, but they got me out of there before I could enjoy a box supper that evening. (Undated interview, 1979)

The arrival of the Air Corps rescue team summarily deprived the young lieutenant of an opportunity to relish the adulation of a fascinated public as well as the company of "a cute young chick." Yet there remained the friendship of the kindly rancher who came to the rescue in a pickup truck. Another Air Corps pioneer, Gen. P. H. Robey, explained that these unexpected friendships became the blessing of many a young airman's adversity. "Many old ranchers at isolated ranches all over Texas always welcomed our Air Force [Corps] types. It was a camaraderie that grew up with the era of forced landings [and] help to each other." Most people living in rural areas of Texas in the early 1930s had no radios, seldom saw daily newspapers, and always enjoyed visitors. Robey and many of his Air Corps friends went prepared to reward those who befriended them. "We always brought along the daily paper and a couple of recent magazines" on cross-country flights, and "some life-long friendships were established" from these unexpected encounters (correspondence, November 14, 1978).

Following the transfer of the Primary Flying School to Randolph Field in 1931, that field soon began clearing flights to Johnson's Ranch, usually in groups of two or more aircraft. According to Gen. Hugh A. Parker, then a Randolph Field instructor, this was a safety precaution issued by the Air Corps training command. He remembered they "usually flew down there in pairs. If you went down [crashed or had a forced landing] in that country with no radio or nothing, you'd be in kinda bad way" (interview, February 8, 1978). By flying in pairs, the pilot of the remaining aircraft could report the location of the downed plane and dispatch help for the grounded crew.

Some of the pilots who flew regularly to the Big Bend airfield did not share Lieutenant Egan's concern for maintaining a safe altitude. Instead, they found hedge-hopping and wilderness acrobatics more to their liking. On one trip to Johnson's Ranch, Egan flew in a three-plane formation led by Lt. Thad Foster. On that occasion Foster flew so low that not only Egan but some Big Bend wildlife feared for their safety. Egan remembered when they flew over the Santiago Mountains between Dryden and Boquillas that "we were just over the tops of the brush and there was a black bear down there that was running like hell to get out of the way." After landing at Johnson's Ranch, the flight continued on to Fort D. A. Russell. While flying over that expansive prairie land south of Marfa they encountered a herd of about twenty-five antelope. Foster immediately gave chase with the three-plane formation, flying at deck level. "They ran ahead of us quite a ways," Egan explained, "and when we caught up with them they turned off. They were really movin' along, must have been going seventy-five miles an hour" (interview, February 7, 1978).

Acrobatics were an integral part of the Air Corps cadet training program, especially that portion of advanced schooling conducted at Kelly Field. Loops, rolls, stalls, and spins were considered essential maneuvers in developing a mastery of the aircraft. These as well as specific combat tactics employed in the pursuit (fighter) and attack (low-level) squadrons were taught, practiced, and executed solely as part of the training program and not for the pilots' or spectators' entertainment.

Colonel Egan regarded the daredevil syndrome as largely a civilian concept based on the poor safety record of civilian stunt pilots. These aerial entertainers achieved wide popularity at county fairs and air shows during the 1920s and 1930s. "Students were not supposed to take unnecessary risks in their training flights," Egan explained. "Maybe they broke the rules once in a while, but on the whole most pilots were pretty well disciplined in the military" (ibid.). Egan obviously spoke in generalities. And while most pilots were well disciplined most of the time, some of Elmo Johnson's regular Air Corps visitors discovered physical challenges in the Big Bend country that led to some rule breaking, hazardous maneuvers that were purposely omitted from their acrobatic training regimen.

One of the constant lures for the Air Corps pilots was a distinctive Big Bend landmark that came into view as they made their final approach into Johnson's Ranch. About five miles off to their right stood Mule Ear Peaks. Appropriately named, two stately rock protrusions stand some five hundred feet above the crest of the mountain, separated at the top by approximately three hundred feet of inviting yet intimidating air space. And to add to the challenge, the opening becomes progressively narrower near the crest of the mountain. In addition, unpredictable wind currents that shift frequently between the twin rock protrusions further increase the hazardous lure to which many bold Air Corps pilots succumbed. Col. Charles Deerwester, who first flew to Johnson's Ranch in a primary trainer on May 16, 1930, claimed to be the first pilot to "thread the needle at Mule Ear Peaks" (interview, February 11, 1971). General Parker never fell victim to the temptation of Mule Ear Peaks, although he flew over it many times, gauging the clearance between the twin projections. He later admitted he "never went down in between 'em. It was a kinda close shave" (interview, April 13, 1978).

Another test of a pilot's skill and courage is found about twenty miles above Johnson's Ranch near Castolon. At that point the dark chasm of Santa Elena Canyon appears to burst through the side of Mesa de Anguila, splitting the rock precipice vertically for more than 1,500 feet. Although the south portion of the canyon is considerably wider than the space on Mule Ear Peaks, the unseen dangers hidden within the bending gorge were sufficient to ward off most Air Corps challengers. Not so Parker. After avoiding Mule Ear Peaks, he tried several times, unsuccessfully, to tame the Santa Elena. "I used to start down that canyon . . . but I never made it. Some of the turns in there were awful close," he recalled. As the canyon narrowed, Parker was forced to roll his BT-2 over on its side to avoid striking the canyon walls with his wing tips. In recovering from this maneuver, he "came out with a lot of rudder; slid out on top. Never made it through" (ibid.).

The Santa Elena also challenged the flying skill of Lt. William L. Kennedy. After watching Lts. Charles Deerwester and B. A. ("Bunk") Bridget run the canyon course in Consolidated PT-3s, he gave chase in a Thomas-Morse O-19. Kennedy vividly recalled seeing "those orange wings winding through that chasm, and [I] thought, hell, if they can fly through that canyon, so can I!" He failed, however, to consider the difference in the maneuverability of the two types of aircraft until he reached the serpentine turns near the eastern end of the canyon. Attempting tight turns at low speeds, the controls of the O-19 suddenly became sluggish and unresponsive; disaster seemed imminent. Even flying at one thousand feet above the floor of the canyon, a parachute would offer little protection in that confined space. "My God, I was scared!" Kennedy exclaimed. "That's how I learned that a PT can outmaneuver an O-19." Kennedy, like Parker, accepted defeat, seeking safety at the last moment by climbing his aircraft above the walls of the winding canyon. He never returned for a second attempt (interview, July 30, 1981).

Thad Foster also challenged the Santa Elena, but for an entirely different reason. Like so many other pilots who visited the Big Bend, the region's raw, primitive beauty fascinated Foster, who took every opportunity to promote the area as a recreation center. On November 19, 1934, he and Capt. Bill Lanagan flew two newsreel cameramen to Johnson's Ranch to film the area, then virtually unknown to the outside world. A Mr. Britton, representing Hearst-Metrotone News, accompanied Foster, and a Mr. Orr, with Fox Movietone News, flew with Lanagan. Their prime objective was to capture the spectacular beauty of the region by filming an aerial sequence of the interior of Santa Elena Canyon.

Neither cameraman, however, volunteered to join Foster for the dive between the canyon walls. Appropriately, one of Foster's previous aerial companions to Johnson's Ranch, Miss Lois Neville (Kelley), expressed complete confidence in the young lieutenant's airmanship. She agreed to serve as the camera operator during the filming of the forbidding chasm. "Thad Foster was a fantastic pilot," she later wrote. "The plane was an extension of himself, and those little planes were amazingly agile and responsive" (correspondence, October 24, 1980).

With the motion picture camera positioned between the two cockpits of Foster's Douglas BT-2, he and Miss Neville strapped on their parachutes and prepared to take off. The plan was to fly through

> Santa Elena Canyon while the movies were being made. The plane was rigged up so that a twine [cord] was passed through from the pilot's front compartment to the passenger's seat behind. . . . Thad Foster was in the front seat, and I was installed with the cord in my hand which I was supposed to pull just as he dove into the canyon. I did. But it [the cord] had gotten caught in the mechanism and broke. . . . The dive pictures did not come out as planned. (Ibid.)

After being defeated by the Santa Elena, Foster elected to challenge Mule Ear Peaks. They continued on eastward toward that famous Big Bend landmark, followed closely by Captain Lanagan and Britton in the other aircraft. As they sped toward the threatening peak, Britton snapped a still photograph of Foster maneuvering the plane into position to "thread the needle." Moments before he and Miss Neville passed between the twin pinnacles, Lanagan veered his plane to the left, avoiding the narrow passage, and followed Foster back to the landing field. The photograph that Britton shot during that daring

Lt. Thad V. Foster, accompanied by Lois Neville (Kelley), emerges from Santa Elena Canyon in a Douglas BT-2 following the filmmaking venture. Kelley remembered: "Foster had repeatedly dived into the canyon with little room to spare for the plane between the rock walls. . . . Very deep in the canyon he maneuvered his plane around and thru the bend seen here at the [eastern mouth] of the canyon." Courtesy Lois Neville Kelley.

Lt. Thad V. Foster in the front cockpit with Lois Neville (Kelley) on the morning of the filmmaking dive into Santa Elena Canyon. Note movie camera on fuselage of the Douglas BT-2. Courtesy Lois Neville Kelley.

filming sequence, though understandably of poor quality, is extant; no record remains of the motion picture footage taken that day (ibid.).

Many Air Corps pilots, according to their personal recollections, engaged in a wide variety of aerial high jinks while visiting Johnson's Ranch. They buzzed ranchers en route, strafed the Rio Grande with machine gun fire, challenged Mule Ear Peaks and Santa Elena Canyon in fragile biplanes, and some even blasted away at high-flying eagles while "riding shotgun" in the rear cockpit of Air Corps observation planes. Their training undoubtedly served them well; no accidents resulted from the antics of those who fell prey to the daredevil syndrome. The one blemish to an almost perfect safety record is recorded with dramatic brevity in the field register—"First Crash at Field." This tragic accident stemmed from neither bold nor ill-advised maneuvers. It resulted from a combination of factors that exacted a high toll in lives and equipment during this period in aviation history—temporary lapses in good judgment and uncertain weather conditions.

The drama began at Brooks Field on the morning of September 22, 1932. Lt. William L. Kennedy emerged from Base Operations after receiving his clearance for a one-day cross-country flight to Johnson's Ranch. A quick check of the latest weather information indicated nothing to alter his plans. With a forecast for local thundershowers and an afternoon high in the upper eighties, conditions seemed ideal for the flight to the Big Bend country. Leaving Base Operations, Kennedy and his passenger, a Private Davis, proceeded to the waiting Thomas-Morse O-19, slipped into their parachute harnesses, and prepared to take off. (This episode was described by Kennedy during an interview on January 19, 1978.)

Once airborne Kennedy executed a routine flight procedure. Seeing the Southern Pacific railroad tracks off to his right, he began maneuvering the O-19 into position for the first segment of the flight. As he climbed to his cruising altitude above that familiar landmark, Kennedy passed through thin patches of summertime cumulus clouds that were gradually being burned away by the sun—typical flying weather for that time of year.

After passing over his usual checkpoints, Kennedy swung north of Del Rio and continued on to Dryden, which he also passed over. He planned to stop there for fuel on the return flight late that afternoon. Leaving Dryden behind, Kennedy set a new compass heading on a southwesterly course and began watching ahead for the Little Dove Mountains. The air was now crisp and clear, the cumulus cloud formations had dissipated, and although the broken terrain created some light clear-air turbulence, it still seemed like a great day for flying. As Kennedy looked ahead over the engine cowling, it appeared that the blue sky must stretch on forever. Unfortunately, that was not the case.

Kennedy will long remember Thursday, September 22, as a day in his life when he became the victim of a convergence of events over which he exercised little control. First there was the early morning weather report. Either the aviation weather center lacked the complete data for the entire state or the staff failed to apprise him of changing weather conditions he was likely to encounter later that day. In any event, Kennedy and his passenger were flying into an area that by sundown would be whipped by violent thunderstorms being pushed ahead of a fast-moving cold front. Before Kennedy left Brooks Field, the weather map indicated that this weather system already stretched across much of West Texas (*San Antonio Express*, September 22–24, 1932). Kennedy, oblivious of the conditions that lay ahead, continued on his flight to Johnson's Ranch. He circled the Chisos range, altered his course to a northwesterly heading, and shortly before noon landed at the Big Bend airfield.

During the course of Kennedy's stopover at the airfield, Johnson mentioned that he had some personal business up at Study Butte and asked Kennedy to fly him there in the O-19. As Johnson was authorized to fly in military aircraft, Kennedy agreed. Sometime around midafternoon they took off for the twenty-minute flight to Study Butte. This was the first of several decisions that Kennedy would make over the next few hours that would greatly affect the course of coming events.

What Kennedy assumed would be a brief visit between Johnson and his friend evolved into something entirely different, another factor that would further complicate Kennedy's flight plan. While Johnson talked, Kennedy began noticing the towering masses of cumulus clouds across the northwestern horizon. At that point he began urging Johnson to terminate his visit, but to no avail. "You know how those West Texas people are," Kennedy explained. "My God, when they start talking, they keep talking. And I was in a hurry to leave, because by then I knew the weather wasn't too good and I was figuring on coming back [to Brooks Field] in the night." Johnson eventually completed his business at the mining village, and as he and Kennedy returned to the parked airplane, the young Air Corps officer made a frightening discovery: the plane had a flat tire apparently punctured by a mesquite thorn during the landing at Study Butte. And to make Kennedy's predicament even worse, the O-19 carried no spare wheel.

At this point Kennedy made an unwise decision. Disregarding Air Corps regulations forbidding temporary tire repairs on the field, Kennedy borrowed an automobile tire patch kit and began repairing the punctured tire. "Now, you're not supposed to do that with an airplane tire," Kennedy explained, "*but we had to.* We pumped it up with a hand pump and by that time it was about dark."

Even as he struggled to repair the tire, the weather became a major factor in Kennedy's thinking. With the protective cloud barrier the Chisos Mountains usually afford he assumed there was still time to outrun the weather front if he stayed south of the mountains. He quickly flew Johnson back to the ranch, picked up Private Davis, and took off just as darkness settled over the Big Bend airfield. Being airborne did little to relieve Kennedy's anxieties; weather and fuel still dominated his thinking. In order to make his fuel stop at Dryden, he had to outflank the weather front south of the Chisos Mountains, and at that point the weather appeared to have the advantage. Dryden was still more than an hour's flying time away over some of the most treacherous terrain in the southwestern United States.

As Kennedy cleared the field at Johnson's Ranch he assumed a southeasterly heading, remaining low until he intersected the Rio

Grande. He then turned east and began following the course of the river. The heavy overcast topped by boiling cumulus thunderheads now appeared to be spreading south of the Chisos Mountains, shrouding the region that lay before him. Kennedy remembered that by then "I really didn't think I would get past Dryden; I probably would spend the night there." But even Dryden was beginning to seem like a remote possibility. With cloud cover obscuring his view of the terrain, Kennedy's one remaining alternative was the Rio Grande, "So I got down in the river canyon and followed it to a point where I knew I wasn't far from Dryden."

Finding the auxiliary field had become a priority factor; approximately one hour's fuel remained in the O-19's fuel tanks. At that point Kennedy made a critical decision; he must abandon the canyon and attempt to find Dryden, no matter the risk. As he climbed into the hovering overcast, he encountered turbulence, rain, and utter blackness. Now, lacking any visual reference by which to navigate, he was suddenly confronted with the stark reality that he would never reach Dryden. But unknown to Kennedy, his greatest menace rode with him, just a few feet away. As he had made his way down the Rio Grande, air had been slowly leaking from the hastily repaired tire. That one ill-advised act would haunt him for years to come. The events of the next few minutes would deem it so.

After probing the turbulent overcast for about ten minutes, Kennedy realized he was in deep trouble and reversed his course. At that point he had two choices, and if these failed there was always the third: bail out. He refused to consider abandoning the airplane as long as there remained the possibility of returning to Johnson's Ranch. However, locating that unlighted field during a thunderstorm was highly problematical. Otherwise he would continue on northwest toward Kokernot Flats and attempt a night landing if he could find a suitable area somewhere in the thunderstorm's wake.

After reversing his course, Kennedy realized almost immediately that his options were fast running out. "I turned around thinking I was going to drop back into the canyon, but I couldn't get in it," he explained. "Back then, you had to see something or you were going to run into something." Unable to maneuver his way back into the river canyon, he had lost the one certain passageway to Johnson's Ranch. And as he continued to fly on a general westerly course heading through the overcast, the needle of his fuel gauge conveyed a brief and emphatic message. He would have to find a place to land the O-19 within the next forty-five minutes or abandon the aircraft.

Suddenly Kennedy discovered an unlikely ally—the lightning. As his westward course ran counter to the thunderstorm's movement, he finally reached the backside of the cloud mass. Brilliant electrical flashes silhouetted the mountain peaks against the northern horizon, and as he broke through the scattered overcast he could identify familiar mountain peaks. "I knew that country by then and I had no problem knowing where I was," he recalled with justifiable satisfaction. From that point Kennedy was able to orient his position and altered his course for the return flight to Johnson's Ranch.

But finding an unlighted landing field in an area lacking any ground illumination—no residential lighting, no street lights, and no automobile traffic lights—was not an easy matter. Still navigating by the momentary silhouettes of the Chisos range, Kennedy finally

> got back to where I figured Johnson's house was and I couldn't see a damn thing. I circled two or three times because it had to be there. But I didn't see anything, so I started up toward a place named Kokernot Flats, up between Terlingua and Marfa. It was the only flat place that I knew that was big enough that I might be able to get down.

The decision to continue on to Kokernot Flats was Kennedy's last resort. The fuel gauge indicated he had less than twenty minutes to locate an unknown spot in an unfamiliar area located an indefinite distance somewhere to the northwest.

With no options remaining, the young lieutenant accepted the consequences of his situation. As he checked his maps for a new compass heading, he recalled that the numbing discomfort of his parachute seat no longer seemed objectionable. It might come to that. But at that exact moment, something else happened.

As Kennedy began his turn to the new compass heading, he "saw a light on the ground and it had to be Johnson's. It was his truck and so, when it stopped, that had to be the field." He would learn later that Johnson had heard the unfamiliar sound of an airplane engine at night and knew someone was in trouble. However, the only help he could provide was the dim headlights on his truck. That was all Kennedy needed. With Johnson's truck parked beside the runway, Kennedy could determine the general location of the landing field.

The O-19 had no landing lights, but carried two manganese parachute flares to aid pilots in making emergency landings in unlighted areas. Taking his bearings from the lights on Johnson's truck, Kennedy circled south of the field and released the first flare, "but I couldn't get in. I had pulled it too high, and before I could get into position to land, it burned out." Kennedy circled the field again, maneuvered his aircraft in a position to land, and released his last flare. There was a brief flash of light as Kennedy got a quick glimpse of the field before him. But "just as I was leveling off to land, it [the flare] drifted down into the canyon. . . . [But by then] I could see the lights from Johnson's truck out of the corner of my eye. . . . I knew where I was. In fact, I was over the field."

Watching the gradual approach of the runway, Kennedy slowly reduced power and felt the plane begin losing flying speed. Momentary forms of mesquite trees below him quickly passed; he was over the runway. Almost silently the O-19 settled toward the soft, wet runway. As Kennedy eased back on the control stick to await the touchdown, he recalled feeling a quiet sense of relaxation come over him. He had challenged the elements and lost, but now, at least, he was back at Johnson's Ranch safely on the ground. Not quite!

It was a good landing—two wheels and the tail skid—just like he had been taught in primary flying school. But then something totally unexpected happened. The airplane suddenly lurched crazily forward, first to the left, then to the right. "At that point all hell broke loose," Kennedy exclaimed. The complete undercarriage was wiped out as the airplane slammed onto the runway on its belly. "Somewhere in there— it was all too quick—the prop [propellor] hit the ground and knocked the engine off. It broke loose and came back over the wing behind me." The wings began buckling as the aircraft ground itself into a mass of twisted rubble as it slid along the runway.

Kennedy remained lucid throughout the crash. As soon as the plane slid to a stop, his first thought was fire. He quickly extricated himself from the wreckage and ran a short distance, but, thinking his passenger might be injured, he returned to the crash site. Davis, still strapped in his seat, looked over the side of the wrecked aircraft and asked, "'Lieutenant, did you cut the switches?' I looked at that mess and wondered, what the hell he thought was left to cut the switches to!" (Following an airplane accident, turning off the ignition switch is a precaution against fire.)

As Kennedy attempted to survey the wreckage in the dark, Johnson drove up in his truck to see if anyone was injured. "He had seen it all," Kennedy explained, "and was very perturbed but pleased that we were not hurt. And he had some dandelion wine that really tasted good." Kennedy's first concern was to report the accident to Brooks Field, but with the nearest telephone located at Castolon sixteen miles away, that would have to wait until the morning.

Kennedy slept very little that night. He arose at daybreak to reexamine the crash site. Retracing the course of the ill-fated aircraft down the runway, it was soon obvious what turned a good landing into a disaster. Etched in the soft wet dirt of the runway were the results of his illicit tire repair. Kennedy discovered the aircraft had actually touched down at three points [a good landing posture] . . . but it didn't stay on three points very long. . . . What happened was that tire—that flat tire—that side started digging in [the soft runway] just as soon as I touched down. Just like a big plow. Then it broke off and the other one went . . . and we ended up on the belly.

The thunderstorm Kennedy had survived the previous night continued to plague him as he and Johnson made their way to Castolon to report the crash. It had left in its wake a wide swath of destruction; they found every bridge and low-water crossing impassable without repairs. Kennedy explained, "We left at sun up. Remember it was only [sixteen] miles to Castolon. We didn't get there until two-thirty in the afternoon. We had to build road all day." Kennedy's call to Brooks Field Base Operations relieved his anxiety but deflated any ego the young officer may have retained. He was shocked to learn that he had not been reported missing. Base Operations had assumed he had decided to wait out the thunderstorm at Johnson's Ranch and would return when the weather cleared. "I had survived an airplane crash and nobody missed me," Kennedy recalled. "That cut me down to size."

Kennedy and Johnson returned to the ranch late Friday afternoon to await the arrival of a team of Air Corps investigators from Brooks Field. Adverse weather also delayed this procedure. It was not until Monday, September 26, that Capt. W. S. ("Pop") Gravely, operations officer with the Twelfth Observation Squadron at Brooks Field, and Lt. Ike Ott, the squadron engineering officer, arrived at Johnson's Ranch in two O-19s, accompanied by Sergeants Smith and Snave. They examined the crash site and the remains of Kennedy's airplane. After questioning Kennedy about the circumstances of the accident, he, the two officers, and one of the sergeants returned to Brooks field later that day. The other sergeant remained at Johnson's Ranch to await the arrival of the salvage team. On October 3, Lieutenant Ott returned to the Big Bend airfield in a C-7A Fokker trimotor accompanied by Captain Scott, Sergeant Hawkins, and five enlisted men "to clean up the mess and complete the investigation" (interview, January 7, 1979).[1]

The loss of an aircraft, due in part to pilot error, might have terminated the career of a second lieutenant less committed to his profession. Kennedy, however, survived the investigation and became one of the United States Air Force's distinguished generals. Yet when re-

---

1. Their work was less than thorough. While visiting the airfield site years later, Kennedy recovered a crumpled portion of a corrugated aileron (he later gave it to me).

calling the incident he expressed mixed feelings about the outcome of the investigation. "When they did the analysis of the accident," he explained, "they called it pilot error, and it was. But I still say that thorn we picked up over at Study Butte had something to do with it."

The significance of the crash of the Thomas-Morse O-19 at Johnson's Ranch lies not solely with the mesquite thorn, the tire repair, the thunderstorm, and certainly not with Kennedy's skill as a pilot. His ability to orient his position during a nighttime thunderstorm, navigate a return course by dead reckoning, and execute a night landing procedure guided only by the dim headlights of Elmo Johnson's truck stands without question. The investigation board's conclusion that pilot error caused the crash—other than possibly referring to the tire repair—is, therefore, both tentative and misleading. A complex of factors obviously contributed to the crash, and the investigating board, in all probability, weighed them judiciously. Yet the one ingredient omitted from the investigation is perspective. When examining this incident, members of the panel undoubtedly ignored one salient fact about military aviation within their own time frame. Regardless of the training, the skill, the precautions, and even the aircraft, flying in 1932 was an unusually hazardous undertaking, carrying with it a high potential for disaster. The significance of the crash of Kennedy's O-19 is, quite simply, *it happened.*

"... order of the white scarf"

CHAPTER 12

As the decade reached midpoint, growing political unrest in Mexico again threatened the tenuous calm that had brought a measure of peace and stability to the borderlands community. In the interim following the retirement of Gen. Plutarco Elías Calles, a disturbing new element had gained prominence in Mexican politics—Marxism. The Communist party of Mexico was organized in 1919; however, during the mid 1930s Marxism became politicized, visible, and influential in both the national party (Partido Nacional Revolucionario, or PNR, the political party that the conservative Calles had organized in 1929) and in the labor movement. By 1934, when Lázaro Cárdenas was elected president of Mexico, left-wing groups had gained control of the PNR. This unhappy alliance led to a breach between Mexico's two current strong men. The ensuing confrontation contained the seeds of future conflict.

Following Calles' departure from Mexico for medical treatment in the United States, Cárdenas embarked on an independent course toward social reform. He ordered the expropriation of lands to peasant villages and encouraged labor groups to assert their rights through strikes. Predictably, this action united Mexico's landed and business elements in opposition to "this sudden deepening of the Revolution. A by-product of this rightist resistance was the Fascist 'Gold Shirt' organization . . . whose members began to clash with Labor and agrarian forces in the streets of the capital" (Lieuwen, 1968: 115).

Calles returned to Mexico in May 1935 and subsequently issued his "patriotic declaration," condemning "labor strikes and agrarian violence and warn[ing] Cárdenas to clamp down lest Mexico's future be endangered" (ibid.). The old revolutionist's reappearance touched off a two-phase, six-month internal power struggle that once again threatened to destroy Mexico's political stability.

Those who manned the ports of entry on the United States side of the border carefully monitored the maneuvering within the Mexican government and began outlining contingency plans in the event the philosophical debate erupted into open rebellion. On June 13, 1935, C. C. Wilmoth, El Paso district director of the Immigration and Naturalization Service, warned the commissioner in Washington of the likelihood of civil war in Mexico and suggested that the appropriate governmental departments begin immediately devising plans for such an eventuality (RG 85). Wilmoth's preplanning went for naught. Calles had misread the signs of the times; Mexico's political center of gravity was shifting to the left. Lacking the support he once enjoyed, Calles announced his political retirement, and on June 19, 1935, boarded a plane for Mazatlán, Sinaloa.

Calles' rejection touched off renewed political dissension, much of it centered in his home state of Sonora. On the same day Calles left for Mazatlán, Walter F. Miller, the veteran inspector in charge at the Nogales, Arizona, port of entry, directed two confidential memoranda to his superior in El Paso, C. C. Wilmoth. One reported a smoldering rebellion in the Nogales, Sonora, garrison between Cardenista and Callista sympathizers; the other described an attack, largely verbal, on an American newswoman visiting the same Mexican border town. The tone of the encounter "was distinctly anti-American and pro-communist, being directed mainly against the capitalist government of the United States" (Miller, June 19, 1935, RG 85).

This emergency passed without any overt conflict. But with Calles' subsequent return to Mexico's political scene—contrary to his premature retirement pronouncement—another crisis was in the making. His appearance in Mexico City December 13, 1935, reignited a storm of protest. As he publicly defended his policies in opposition to the liberal Cárdenas regime, it became readily apparent that the hero of the Battle of Jiménez had lost touch with the political realities of Mexico. His weakening political base failed to sustain him; on April 10, 1936, the last member of "El Triángulo Sonorense" was banished from Mexico.

Mexico City and Nogales are far removed in both time and distance from Johnson's Ranch. And while there is no record of local conflicts between the Cardenistas and the Callistas in the region of Mexico adjacent to the Big Bend airfield, the impact of this political upheaval ultimately made itself felt north of the Rio Grande. The nationalization of Mexican industry owned by foreign investors, the oil industry especially, heightened the diplomatic tension between the United States and Mexico during the latter half of the decade. This opposition to foreign investment found support at both extremes of the political spectrum. On the far right stood the Gold Shirts and the Sinarquistas, while to the left the growing influence of Marxism within the Mexican government further widened the philosophical breach between the two countries that meet at the Rio Grande.

And while Brewster County judge R. B. Slight may or may not have been aware of these philosophical differences, fear, distrust, and the imagination of what existed beyond the Rio Grande remained a powerful influence among borderlands citizens. Judge Slight recognized within the Escobar Syndrome a negotiable currency that he could use to support his request for Works Progress Administration funds to expand the public-owned facilities at Johnson's Ranch. (The Johnsons had deeded ten acres of land containing the officers quarters and the radio station to Brewster County.) In his communication dated October 12, 1935, Judge Slight explained to Ben Boykin, acting chief examiner for the Works Progress Administration in El Paso, that the airfield was located in

> the most vulnerable spot along the Border between Presidio and Sanderson. For that reason, we have been endeavoring to make a place available for Officers to concentrate and stop in that isolated section.
>
> The Border Patrol service, that portion which uses airplanes, requested and obtained through us, by considerable expense, an emergency landing field at this point. After landing, there is no place for the pilots and passengers of such planes to get out of the weather or to be served with the necessities of life.
>
> For the protection of the citizens in that isolated portion of this country, and to accommodate the local County authorities, members of the Ranger force, the Immigration service, the Border Patrol service, and other persons on public business, we are endeavoring to build [temporary living] quarters. We already have partially completed a building which houses a radio station, the equipment and operators of which are furnished by the Federal Government on a twenty-four hour basis. (Slight, October 12, 1935, RG 237)

The image of Gen. José Gonzalo Escobar still lingered beneath that great cottonwood tree across from Johnson's trading post. But by 1935 it was only a shadow.

Knowingly or otherwise, Judge Slight obviously strayed widely from the facts in pleading his case to Boykin. According to Air Corps pilots who frequented the Big Bend airfield, the facility known as the "officers quarters" never fulfilled the judge's expectations. While the military pilots were welcome to stay in the Johnson residence, most preferred to sleep outdoors. And if the weather was unfavorable for outdoor activities, they stayed in San Antonio, as Col. John W. Egan explained. Also, the emergency landing field Judge Slight refers to was established by the United States Army Air Corps and Elmo Johnson, and was not "obtained through us [Brewster County], by considerable expense." Nor did this facility ever become a major stopover for the airborne Border Patrol. According to Chief Patrol Inspector George W. Harrison, "We had to call El Paso if we needed the patrol plane as there was only one in the district. I never knew of it landing at Johnson's Ranch. I never met Johnson; he had no love for the Border Patrol" (correspondence, September 26, 1980).

Judge Slight was, nevertheless, successful in his effort. Brewster County received $10,086.72 in federal funds to "provide quarters for air pilots, border patrol, customs officers, game wardens, and any public official or government employed men whose services are required at this isolated frontier point" (WPA Forms 301 and 306, RG 237).

Although the Calles-Cárdenas crisis in government touched some nerve centers along the Arizona-Sonora border as well as at the diplomatic strongholds in Washington and Mexico City, it did little locally (other than to give Judge Slight some verbal ammunition) to disturb the usually quiet solitude of the lower Big Bend country. With organized border disturbances no longer an impending threat, Big Bend security remained the part-time responsibility of the United States Army Air Corps operating out of Johnson's Ranch. This became a seemingly vacant mission. Although combat aircraft appeared at this remote station with increased frequency, no shots were ever fired in anger. To the Air Corps pilots who regularly visited Johnson's Ranch, the reason was soon apparent; by mid decade the river merely separated good neighbors.

The cursory impression was, in the main, valid. Most people, both Mexican and Anglo, who populated that area adjacent to the Rio Grande lived harmoniously with one another. There remained a shadowy criminal minority, protected by that isolated stretch of the international border, who continued to prey on the Big Bend ranches throughout most of the decade. Although the periodic loss of horses, cattle, and goats became more of an annoyance than a threat to personal security, Elmo Johnson and his neighbors still regarded the appearance of military aircraft as their best protection against border depredations.

Between July 11, 1933, and June 27, 1934, the field register showed a slight decline in air traffic from the previous year: forty-eight aircraft bearing eighty-five personnel. All traffic originated from the San Antonio–area bases (Randolph, Brooks, Kelly, and Duncan fields) and Fort Crockett. The following year, the sixth year of operation, air traffic almost doubled: ninety-two landings and 154 service personnel were reported at the airfield.

The field register for this period also reflected the rapid technical advances in the field of United States military aviation. During the 1934–35 fiscal year, five new aircraft types appeared at the Big Bend airfield. In addition to two Curtiss-Wright observation-type aircraft, the O-40 and the O-43, whose military service was short-lived, a new attack, a new observation, and a revolutionary new bomber, representing the most advanced combat aircraft in the United States if not the world, appeared at Johnson's Ranch.

On September 14, 1934, Lts. John W. Egan and Robert S. Macrum, accompanied by Sergeant Todd, arrived at the Big Bend airfield in the Air Corps' latest long-range reconnaissance aircraft, the Douglas O-35 (also designated the B-7 when assigned to light bombardment squadrons). When the Air Corps accepted this unique twin-engine, high-wing monoplane for field testing, it was faster than most of the Air Corps' fastest pursuit types and had a longer range than any bomber then in service. Powered by two six-hundred-horsepower, liquid-cooled Curtiss "Conqueror" engines, the O-35 attained a maximum speed of 182 miles per hour. The retractable landing gear and the all-metal fuselage represented engineering milestones in the development of high-performance aircraft.

The aircraft that ultimately rendered the O-35 obsolete, revolutionizing bombardment aviation as well as altering the tactical concepts of pursuit aviation, arrived at Johnson's Ranch five days before Lieutenant Egan landed the O-35. The brief field register entry— "9/9/34 B-10 Capt. John Corkle [sic] Alaskan Flight"—belies the significance of the occasion. This marks the appearance of not only another revolutionary aircraft design, but the only "famous" aircraft to be registered at the Big Bend airfield.

The Glenn L. Martin Company of Baltimore, Maryland, received the 1932 Collier Trophy for developing the B-10 bomber. This award, given annually, recognized that year's outstanding contribution to aviation progress. Two 675-horsepower air-cooled Wright "Cyclone" engines powered the all-metal, mid-wing monoplane that weighed 12,829 pounds fully loaded. With a maximum speed of 207 miles per hour, the B-10 was faster than the Air Corps' fastest pursuit plane, yet could carry a 2,260-pound bomb load, the largest of any bomber manufactured at that time.

The nation became aware of this aircraft when Lt. Col. Henry A. ("Hap") Arnold led a formation flight of ten Martin B-10 bombers from Bolling Field, Virginia, to Fairbanks, Alaska. Fourteen officers and nineteen enlisted men embarked on this 7,335-mile mission on

July 20, 1934, described by the War Department as a training and photographic flight. With this unprecedented achievement, Arnold demonstrated the Air Corps' long-range bombing capacity with both speed and endurance. This feat typified the Air Corps' continuing struggle for service status and popular recognition at a time when speed dashes, altitude records, and long-distance endurance flights fascinated an aviation-conscious public. Equally impressed was the United States Congress. Following the successful field testing of the B-10s, the Air Corps received an appropriation of eighty additional Martin bombers.

Capt. John D. Corkille served as flight engineering officer and pilot of aircraft No. 150 on the Alaskan flight. No one recalls why he included Johnson's Ranch on his itinerary after returning to the United States. A sentimental journey is a fair assumption. He first visited Johnson's Ranch on December 8, 1933, when he led his Eighth Attack Squadron from Fort Crockett on their first Big Bend "field maneuver."[1] Corkille returned to Johnson's Ranch the following November 15, again leading his Eighth Attack Squadron on its second annual "field maneuver." The squadron arrived in fourteen new Curtiss A-12 "Shrike" attack planes and one Y1-C cargo plane, the largest single concentration of military aircraft ever assembled at Johnson's Ranch.

The A-12, which replaced the venerable A-3 biplane, incorporated advance design features that accorded this low-wing, all-metal attack plane an important niche in the history of military aviation. Aviation historian Jack Dean views the "Shrike" series as symbolic of this transitory period in aviation history when technological advances were fast broadening the field of combat aviation. He writes:

> The period 1930–1933, was a pivotal point in aircraft design in this country and throughout the world. The day of the all metal monoplane was dawning; the era of the fabric covered biplane was drawing to a close. True, the latter would be around at the end of the decade, but few, if any, planners thought in terms of biplanes when designing for the Army Air Corps' new criteria: speed and narrow frontal area to enhance streamlining. (Dean, 1979: 3)

The A-12 far surpassed its immediate predecessor in performance. A 670-horsepower Wright "Cyclone" engine gave the two-place, low-

---

1. Corkille and Lt. P. H. Robey, another member of the Eighth Attack Squadron, were later reunited to help develop another "famous airplane." They helped conduct the original tests on the supercharged version of the Boeing B-17 "Flying Fortress" just prior to World War II (Robey correspondence, November 15, 1980).

level attack craft a maximum speed of 177 miles per hour and an operational range of 510 miles. The large circular engine cowling and the
massive streamlined wheel "boots" cast a silhouette of sinister brutality.
Five .30 caliber machine guns—four forward and one rear swivel gun—
plus various bomb configurations confirmed the A-12's visual impression. The A-12 also incorporated two significant aerodynamic innova-
tions. Trailing edge flaps (then called speed brakes) permitted slower
landing speeds, and leading edge wing slots gave added lift and stability to the aircraft. These two design features would appear on many
subsequent aircraft types, both commercial and military.

In a retrospective view of attack aviation, this unique airplane
stands at the pivotal juncture between the fabric-covered A-3 biplane
and the twin-engine medium bombers that performed such critical
roles in every combat theater during World War II. The pilots who
negotiated the some five hundred miles between Fort Crockett and
Johnson's Ranch in the innovative A-12 "Shrike" unknowingly were
sharpening their multiple skills for a conflict that still lay almost a decade in the future.

Reflecting on the squadron's peacetime junket to Johnson's Ranch,
Col. Fred R. Freyer recalled that "we were fortunate in always having
a CO [commanding officer] who believed in authorizing diversions,
such as duck hunting and fishing (Matagorda Peninsula) and deer
hunting (Johnson's Ranch). Normally some justification—military
related—was necessary; hence, Field Maneuvers" (correspondence,
October 10, 1978). In describing the squadron's week-long sojourn
with the Johnsons, Freyer reiterated a recurring theme articulated by
many Air Corps pilots who visited that remote facility: the deployment of military aircraft as a threat to border incursions. For this
reason, Johnson always extended a cordial welcome to all military
pilots as

> the Mexicans foraged across the Rio Grande and usually returned
> home with too many of his livestock. Therefore, a show of power
> in the form of military aircraft was intended as a deterrent to this
> rustling. The fact that Mr. Johnson was a cordial host and made
> available hunting privileges with respect to the deer on his domain
> was incentive enough to interest the more avid nimrods among
> the Air Corps officers. (Ibid.)

To the outdoorsmen of the Eighth Attack Squadron, Johnson was
the ultimate host, supplying burros, a wagon drawn by six mules, and a
guide to transport the visiting airmen to the more productive "hunting
grounds in the far away mountains." But according to Freyer, all was in
vain, as "our intrepid huntsmen, all bedraggled and worn out, re-

turned to the base camp and reported that the only 'deer' shot, turned out to be one of Mr. Johnson's goats! We dined on 'cabrito' that night!" (ibid.). Much of Johnson's accommodations, Freyer recalled, consisted of the great outdoors, as the Air Corps visitors

all slept under the stars on GI cots, and the mess facilities and food were a far cry from those to which we had become accustomed. . . . the food was plentiful and wholesome, and was prepared by the enlisted men who were brought along for that purpose. We ate from typical army mess kits and dined rather unceremoniously in the open. (Ibid.)

Freyer enjoyed the Big Bend outing, but, not being a hunter, he

remained in the base camp area, and interested myself by taking hikes along the banks of the Rio Grande, and enjoying the utter wilderness of it all. . . . There was a wild beauty about that Big Bend area, and I remember vividly the sometimes purplish hue of the distant mountains. . . . I had never seen anything like it before. (Ibid.)

In retrospect, Freyer regarded the squadron's November odyssey objectively, concluding that "it is questionable that the unit gained much from a purely military standpoint. It was more of a fun thing" (ibid.). Gen. P. H. Robey, who participated in the squadron's previous "field exercise," was even more candid. Pointing to the apparent margin between duty and recreation, he acknowledged, "There may have been some higher HQ [headquarters] contacts [between Johnson's Ranch and Fort Sam Houston], but I would suspect that this merely poured the holy water on an excuse to go hunting" (correspondence, November 14, 1978).

Freyer's and Robey's comments on the use of military aircraft at Johnson's Ranch for personal recreation also apply to the current status of military aviation in the 1930s. The relaxed informality that characterized the Air Corps' Big Bend operation reflected to a marked degree the operational policy of the entire service. This condition evolved largely in accordance with the origin, size, combat record, and traditions of the United States Army's aerial combat force. It also partially explains the continued use of Johnson's Ranch as a quasi-private, quasi-military, and totally remote haven for many Air Corps personnel.

The United States Army Air Corps evolved as the junior member of the nation's defense establishment. Only eighteen years separated the

Eighth Attack Squadron's autumn sojourn and the United States Army's first tactical use of aircraft.[2] The First Aero Squadron's departure with the Pershing Punitive Expedition into Mexico in March 1916 marks the initial use of aircraft by the United States Army in a military operation (Maurer Maurer, 1961: 1–2). The undistinguished record of the fragile Curtiss JN-3 biplanes, used primarily for reconnaissance and carrying dispatches, achieved little to portend the airplane's military potential. But, more important, approximately eight months constituted the entire combat record of United States military aviation at the time of the 1934 "field maneuver" at Johnson's Ranch. According to aviation historian F. G. Swanborough, the first United States aviation unit did not reach Europe during World War I until September 3, 1917, and it was not until the following April 3 that the Ninety-fourth Pursuit Squadron went into action on the Western Front (Swanborough, 1963: 2).

From this brief period of aerial combat the United States Air Service (and later the Army Air Corps) emerged with a romantic tradition based largely on the wartime records of a small cast of international heroes. The names and faces of some became national symbols. For example, the photograph of Capt. Eddie Rickenbacker leaning jauntily against his diminutive French Spad biplane received wide circulation in the nation's magazines and newspapers. His engaging smile, highly polished boots, and countless air medals decorating his expansive chest became symbolic of all World War I aerial heroics. His equally well-publicized off-duty escapades involving champagne, girls, and late nights in Paris bistros further glamorized a heroic tradition that ultimately became part and parcel of military aviation.

Stringent requirements for pilot training during World War I (and even later) reflect the elitest concept of American manhood and the high expectations of those who flew the nation's fighting aircraft. The strict requirements—physical, mental, moral, and psychological— elevated further the standards for candidacy. To be accepted for pilot training,

> The candidate should be naturally athletic and have a reputation for reliability, punctuality and honesty. He should have a cool head in emergencies, good eye for distance, keen ear for familiar sounds, steady head and sound body with plenty of reserve; he should be quick-witted, highly intelligent and tractable. Imma-

---

2. The United States Army established an Aeronautical Division in the Signal Corps on August 1, 1907, acquired the first airplane in 1909, and formed the first First Aero Squadron in Texas on March 5, 1913.

ture, high strung, over-confident, impatient candidates are not desired. . . . One man out of two—18,004 to be exact—was rejected under these criteria. (Glines, 1963: 78)

The survivors became legendary. Ingenious entrepreneurs quickly capitalized on the popularity of America's new idols of the sky, and began packaging their accomplishments—both real and imagined—for public consumption. Millions flocked to the nation's theaters to see such silent films as *Wings*, *Lilac Time*, and *Hell's Angels* (reshot for sound), while another audience relished the printed word. A vast array of pulp magazines—*War Birds*, *Flying Aces*, *Sky Fighters*, etc.—depicted the World War I airman in his most exciting and romantic setting. Costumed in leather helmet and goggles, with the ever-present white scarf streaming behind, the image-makers pictured him challenging the forces of evil alone in what ultimately became "the wild blue yonder." Such are the seeds from which folk heroes grew.

The romantic image spawned in the post–World War I era received an additional boost during the 1930s. Throughout much of the inter-war period the Air Corps encouraged pilots to fly at every opportunity in order to sharpen their skills. Considering the general proficiency of that period, this policy was well advised. Gen. Hugh A. Parker, a second lieutenant with the Third Attack Group in 1931, explained, "We had pilots in that outfit, some were World War I pilots, who still had not flown 500 hours." Most squadron commanders, recognizing the value of practical flying experience, however gained, encouraged their pilots to accumulate additional flying time whenever possible. In explaining the application of this policy, Parker emphasized the personal freedom accorded Air Corps pilots in the early 1930s.

Back in those days they actually encouraged people to fly. We could take a cross-country [flight] just about anytime we wanted to, which was in the interest of improving your pilot proficiency, navigation, landing in strange areas, and what not. . . . If you wanted to go to Dallas and back on the weekend, or to New Orleans, or any place within five hundred miles, you could go. Twice a year, by request, they would give you an extended cross-country, provided you could be spared from your station. This meant you could go anywhere in the continental United States. They were encouraging you to get out and fly and find out what that bird [the airplane] was all about. (Interview, February 8, 1978)

The "get out and fly" policy was always contingent on the availability of fuel and it fluctuated with the time of the fiscal year. Gen.

Robert S. Macrum remembered receiving instructions "along in the spring . . . to conserve on fuel, just get in the minimum amount of hours of flying. . . . Stay locally and do your minimum training." This policy was subject to change. As the end of the fiscal year approached, the squadron commanders frequently received urgent orders from Washington stating that the Air Corps had inventoried an abundance of fuel that must be consumed before the end of the fiscal year to prevent a reduction in the following year's allocation. Macrum recalled one June "practically all airplanes from Brooks Field were all off on extended cross-countrys to try to use up the gas before the end of the year." He received clearance for a flight from Brooks Field to Portsmouth, New Hampshire, and back during the month of June "as part of that [policy of] using up the gasoline that was in storage" (undated interview, 1979).

This policy of encouraging young Air Corps officers to gain independent flying experience was one reason that Johnson's Ranch flourished for more than a decade. "It was acceptable to keep the aircraft [out] overnight and over the weekend," Col. Don Mayhue explained, "provided that the operations officer knew where the pilot and equipment were." But however it may appear in retrospect, "it was not all just fun and games. I was fulfilling an assignment in flying to the Big Bend, but once there, I could relax and have fun" (interview, February 7, 1978).

Under the "get out and fly" policy, many young officers received official flight clearances to the most unlikely destinations and for the most unlikely purposes. These included weekend barbecues and dances at Johnson's Ranch. Such activities seemed to always coincide with their need for additional cross-country flight training or exercises in strange field landings. A mammoth barbecue-fishfry at Johnson's Ranch remains vivid in Gen. Joseph H. Hicks' recollections. It was staged in the best Texas tradition—an elaborate spread executed in a manner that bordered on the extravagant. To fabricate a more colorful and lavish gathering than the one held at Johnson's Ranch on Sunday, June 12, 1932, would challenge the imagination of Edna Ferber. But there the similarities end. Frontier generosity was the motivating force, and these characters were genuine. Some came on horseback, in wagons, and in cars, while others found it more convenient to fly in.

As preparations for the event got under way, the Big Bend air station began to resemble a "gathering of eagles." On Friday, June 10, a Lieutenant Williams and a Sergeant Jones arrived from Randolph Field in a BT-2, and the following day Hicks, accompanied by a Sergeant Martin, landed at the Big Bend airfield in an O-19, followed shortly by Lt. Thad Foster and a Major Coker in another BT-2. On Sunday morning two more O-19s from the Twelfth Observation Squadron at Brooks Field arrived for the festivities. These were flown by Lieutenants Greer and Rawlins, with Privates Martin and Smith as their passengers.

Hicks remembered that everything about the Big Bend social—the airplanes, the food, the crowd, and the preparations—was on a grand scale. Even the catfish he and Johnson caught in the Rio Grande were plentiful and enormous, some measuring from four to five feet long. Hicks explained that the size of the catch was matched only by the size of the crowd. "Where all the people came from, I don't know. . . . When you flew over that country then, you saw very, very few ranches" (correspondence, December 15, 1979).

Dances at the riverside airfield also attracted a host of young aviators from the San Antonio–area bases. The lower Big Bend country, however, lacked two essential ingredients for a successful dance—liquor and women. The Johnsons provided the latter, inviting young ladies from local ranches as well as from Terlingua and Marathon. Securing a supply of liquor during Prohibition was a far more difficult task, even on the banks of the Rio Grande. Gen. Hugh A. Parker explained: "When we'd go down there, we'd take a jug of whiskey with us. It was old Bourne [Texas] whiskey. Three dollars a gallon. It'd knock the top of your head off!" (interview, February 8, 1978). Parker also investigated local sources of Mexican whiskey without success. The main obstacle was Pete Crawford, the state game warden who exercised extralegal authority in that area. Parker even offered to fly the illegal booze up from Boquillas in his Air Corps BT-2 trainer if Crawford's forbearance could be assured. But the game warden's strict adherence to the law annoyed the young lieutenant. "Used to gripe the hell outta me," Parker recalled. "I'd say, 'Pete, if we're going to have these dances, why in the hell don't we go down to Boquillas and bring some whiskey up?' Pete'd say, 'Oh, no! Oh, hell no! If I ever catch you, I'm gonna run you in.'" Parker sat thoughtfully for a moment, considering his old adversary's threat, and added, "And I think [he] would have. . . . He was a tough *hombre*" (ibid.).

Despite the problems of distance, liquor, and women, Parker pronounced the weekend dances at Johnson's Ranch highly successful. He recalled that sometimes Johnson would recruit a local Mexican string orchestra, and "when the fiddlers got tired they would crank up that old Victrola and turn 'er on." The guests danced either in the front yard or "they'd clear out the furniture on the patio . . . [which] was quite large and had a concrete floor on it. We'd really cut a rug, square dances mostly" (ibid.).

The personal use of military aircraft contributed to the pilot's proficiency, and the interest they generated by landing in areas where the airplane was still a novelty conveyed a positive image of the Air Corps. Wherever they went, America stopped, looked, and long remembered. Gen. William L. Kennedy explained the pilots' unique role in the total scheme of things and the benefits the Air Corps gained from their visibility.

> The situation with the old-time airplane was so different that it's not comparable to nowadays. You could land lots of places; I landed and hunted and shot quail around this area [San Antonio]. For example, you just landed and the rancher was glad to see you, you were glad to see him, and the next thing you knew you were friends. And it was the same with the small towns. The trouble with that was that when you landed, my God, everybody that could, would be there in no time to see the airplane. And when it came time to take off, they were all in your way. We didn't have any public relations, but I will say this . . . our public relations were pretty good. (Interview, January 19, 1978)

The individual use of military aircraft, officially sanctioned and encouraged by the Air Corps administration, produced a level of individual freedom among young officers that had no counterpart in other branches of the service. This freedom prompted an admiring public to enthusiastically accord the Air Corps officers a hero's stature, which they may or may not have solicited. When asked if he was cognizant of this public acclaim, Kennedy responded thoughtfully, "Well, I couldn't even carry my parachute from the plane to the hangar at a civilian airport. A bunch of kids would be begging for the privilege" (ibid.). Col. Roy P. Ward's self-awareness of his public image was even more positive.

> Flying was quite a novelty in those days. The stories that came out were World War I fighter [pursuit] pilot stories, and, of course, they were highly romanticized. . . . After we graduated from flying school and got those high tan boots and silver wings on our blouses, you could walk down the street and pick up damn near anybody you wanted to. Hell, I was a hero and a celebrity too! (Interview, April 9, 1980)

Heroes and celebrities, young men in helmets and goggles, flying the airplanes with the distinctive orange wings accented with the red, white, and blue star insignia, these were the bearers of a heroic tradition that spanned three continents and fascinated an admiring public for more than three decades. Eventually the mantle was passed to the hands of the Thad V. Fosters, the Roy P. Wards, the William L. Kennedys, and other officers of this later era. They carried it with style and individuality, and subsequently passed it on to others. The "Order of the White Scarf" was destined to thrive and grow with the United States Army Air Corps.

And while the sanctioned training benefits accorded these young heroes included the personal use of military aircraft, such reasons as navigation training, improving pilot proficiency, and landing in strange areas insured an official clearance to various destinations and for various reasons. These might include a football game in Dallas, a quail hunt near San Antonio, or a day, a night, a weekend, or a week-long "field maneuver" at Johnson's Ranch.

". . . Elmo just asked us to help out and we volunteered"

CHAPTER 13

The continued presence of Air Corps personnel in the Big Bend setting touched the lives of many people in many different ways. Training
missions to the riverside airfield not only added a significant dimension to the pilots' professional as well as personal experiences, but also
enriched the lives of many people—both Anglo and Mexican—living
along both sides of the Rio Grande.

In the process of fulfilling their military commitments, the young
and mostly single airmen, found the cost-free recreational opportunities at Johnson's Ranch appealing, especially during the Depression
years. And in accepting the Johnsons' hospitality they in turn brought
with them a unique complex of skills and services heretofore unavail-
able in that remote section of Texas.

During the Air Corps' fourteen-year tenure at Johnson's Ranch,
members of the Big Bend community responded to changes in their
social life, health care, and especially international relationships.
These aspects of the Big Bend lifestyle assumed new meanings under
the auspices of the Air Corps "get-out-and-fly" policy. While all members of the partnership found the contacts mutually beneficial, the
military visitors' dedicated performances sometimes exceeded the
ethical, legal, and statutory perimeters of acceptability.

While apparently legal, free weekend medical clinics established by
members of the Aviation School of Medicine in Elmo Johnson's Trading Post were regarded by some as only marginally ethical. Giving civilians free Sunday-afternoon rides in military aircraft was categorically
prohibited by Air Corps regulations, even though it was popular with
the local citizenry. And the use of United States military aircraft to
pursue suspected murderers (who were also Mexican nationals) into
Mexico was not only prohibited, but violated every premise of inter-

national diplomacy. But, what the hell! This was Johnson's Ranch deep in the Big Bend country of Texas, and who was going to blow the whistle?

Flight surgeons—Air Corps physicians specializing in aviation medicine—exhibited an early interest in Johnson's Ranch. The first medic to visit the facility, Maj. Keith Simpson, registered at the field on July 5, 1930. Signatures of other members of the Aviation School of Medicine, then based at Brooks Field, appear subsequently in the field register. Maj. C. E. Breen, a flight surgeon attached to the Third Attack Group at Fort Crockett, was also a frequent guest of the Johnsons. When the Air Corps transferred the Aviation School of Medicine to Randolph Field, Breen joined the medical staff there and began flying to Johnson's Ranch from that facility.[1]

The archaeological specimens found in that area also interested Breen, and he encouraged a scientific exploration of the sites identified by Mrs. Johnson. In 1932, Frank M. Setzler led a Smithsonian Institution field party to the lower Big Bend, and at Breen's suggestion headquartered at Johnson's Ranch. As might be expected, the Air Corps cooperated unofficially. Setzler's report states that he surveyed the Chisos Mountains from the air and "numerous caves were sighted along the precipitous cliffs. The territory we covered by plane in an hour required four days to reach by muleback" (Setzler, 1933: 53).

On April 16, 1932, Breen returned to the Big Bend air station in a Douglas BT-2 basic trainer piloted by Randolph Field flight instructor Lt. Sigma Gilkey. On June 5, two more medics, Majors Mike Healy and Keith Simpson, flew to the ranch with Lts. Elmer P. Rose and Hamp Atkinson in the same type aircraft. Simpson's reason for making the trip, in addition to fulfilling his four-hour monthly in-flight requirement, was to examine the skeletal remains of a prehistoric Indian culture Mrs. Johnson had recovered from nearby caves.

The periodic visits of the Air Corps flight surgeons became a matter of great local interest, as people living in the lower Big Bend in the 1930s were virtually without medical care. Emergency cases—mostly Anglo—were taken either to a mining company doctor at Terlingua or a hospital in Alpine. Time and distance rendered both highly im-

---

1. In the early 1930s, the science of aviation medicine, like military aviation itself, was in its formative stages. According to contemporary observers, the young physicians who volunteered for the Air Corps were dedicated to insuring the pilots' physical fitness. According to Gen. William L. Kennedy, "They haunted the operations office, and they were masters of preventive medicine. If there was something wrong with you, you weren't going to fly" (interview, July 30, 1981).

Air Corps visitors relax in the covered patio at Johnson's Ranch. Seated left to right are Lt. George W. Hansen, a Lt. Derosier, Lt. Mark Lewis, unknown, Ada Johnson, Elmo Johnson, Capt. Keith Simpson, a flight surgeon, and unknown. The unidentified officers were students at the Aviation School of Medicine. Indian relics, Smithers' photographs, Eighth Corps Area insignia, and Elmo Johnson's hunting trophies adorn the walls. Courtesy George W. Hansen.

practical, especially for the Mexicans, who depended largely on local *curanderos* for their health care. But the flying physicians at Johnson's Ranch brought new hope for sick Mexican people; the "Mexican telegraph"—the *avisadores*—circulated the good news along both sides of the Rio Grande. According to Col. George W. Hansen, the meeting was mutually rewarding, as "the doctors had a field day every time we went to Johnson's. The MDs were willing passengers, especially Keith Simpson and Charlie Glenn. . . . word soon got out and Mexicans would come in from miles around to get treated" (correspondence, September 18, 1978). While the distant roar of United States aircraft sounded a warning to some living south of the Rio Grande, to others it meant relief from pain and suffering. According to Gen. Hugh A. Parker, "just as soon as they [the Mexican people] would hear those airplanes up and down the river, you could figure in a matter of hours there'd be a bunch of 'em over there." The Air Corps physicians began carrying their medical kits in anticipation of any emergency. Parker explained that Glenn "would pull teeth or anything else that showed up that they wanted done. . . . They treated skin disorders, stomach disorders, dressed wounds, and just about everything that could go wrong" (interview, February 8, 1978).

Gen. William L. Kennedy remembered seeing the Mexican people lined up in front of the trading post awaiting admission to Elmo Johnson's "free clinic." The doctors would minister to all who needed care and frequently left a supply of drugs, medicine, and bandages with the Johnsons along with complete instructions on how they were to be used and administered. "Mrs. Johnson mothered people all over that country," Kennedy explained, "and Elmo, although he was brusque in manner, was just the same" (interview, January 19, 1978).

W. D. Smithers discovered a case for the Air Corps medics while visiting a Mexican family in Mexico. One of the children, an eight-year-old girl, had a crooked left arm; about half-way between the elbow and the wrist, the limb bent abnormally at about a twenty-degree angle. Three years before she had fallen off a burro and had broken her arm. With no doctor to set the fractured limb, the bone regenerated in that distorted form.

Smithers persuaded the parents to bring the child to Johnson's Ranch to be examined by the Air Corps physicians. The diagnosis was immediately apparent; re-break and re-set the arm. Majors Glenn and Simpson agreed to perform the procedure. The first priority was anesthetic; one of the visiting pilots volunteered to fly up to Terlingua, where he acquired a supply of ether from the company doctor employed at the Chisos Mining Company.

The physicians, with Smithers' help, converted the counter of the trading post into an operating table. "We spread a couple of blankets on top of the counter in the store," Smithers explained, "and spread a sheet on top of that, and laid the boy [it was a girl] out on that." He adds that he held the cotton with the anesthetic to the child's nose as the doctors prepared for the operation (interview, August 27, 1975).

Parker, who watched as Glenn began the procedure, said, "Charlie took her arm over the edge of the counter and broke it. You could hear the bones pop. He then re-set it straight, securing it with some splints, tape, and bandages." Substitute material to fabricate the splints presented a temporary problem until Smithers produced some balsa wood planks he used in constructing his homemade cameras. Despite the inauspicious setting and unorthodox procedures, the young patient enjoyed a complete recovery. Parker returned to the ranch a few weeks later and found "the little girl's arm was perfectly normal, back straight" (interview, February 8, 1978).

The flight surgeons' voluntary health care was not limited to the Mexican community. On two occasions the Air Corps' emergency medical service came to the aid of Elmo Johnson; following one accident, the flight surgeon probably saved his life.

General Kennedy remembered receiving an emergency message relayed to the Brooks Field operations office from the Eighth Corps Area Headquarters. His instructions were brief and to the point: "'Elmo is in very, very bad shape. Go down there and get him and let's put him in the hospital.' So I got in the airplane right away and went after him." (Details of this episode, unless otherwise cited, are taken from an interview with Kennedy on January 19, 1978.)

When Kennedy left Brooks Field no one knew the nature of the emergency. After he arrived at the ranch and found Johnson lying on a cot in the patio, questions were superfluous. Nothing distorts the human body like a rattlesnake bite. A half-century later, the sight remains vivid in Kennedy's recollection. "His leg was black, swollen all the way to his hip and had cuts—incisions. That's where he bled it." Kennedy agreed with the medic's long-distance diagnosis; Johnson was in "very, very bad shape."

Johnson's misfortune resulted from his unyielding position on certain household amenities. "Elmo had a fixation about indoor plumbing," Kennedy explained. "He didn't want a toilet in the house." So, on this particular occasion, as Johnson went outside at night to relieve himself before retiring, he violated the unwritten law of the rattlesnake country—don't step where you can't see!

Johnson was wearing house slippers as he stepped into the darkness, and

as he started, he heard the rattle and the reaction was automatic. He kicked and the slipper came off and the snake bit him. One fang hit one toe and one the other. Well, he knew he was in bad trouble. Mrs. Johnson had fever blisters on her mouth and he wouldn't let her draw out the venom, and he couldn't get his foot in his mouth. So all that was left for him to do was to cut the bite and try to let it bleed, and relieve the pressure from the swelling.

The next day Johnson was driven to Terlingua and Alpine, where he consulted two doctors, but he refused to be hospitalized and insisted on returning to the ranch. That was where he intended to stay—live or die—and by the time Kennedy arrived at the ranch, Johnson felt his chances of surviving had improved. Kennedy and Mrs. Johnson were not so optimistic. It was she who had alerted the Eighth Corps Area headquarters to her husband's worsening condition. Considering Johnson's immobility, Kennedy's prime concern was getting him into the open cockpit of the airplane. With his leg so swollen, painful, and rigid, that task appeared hopeless. The young lieutenant's apprehension, however, was ill-founded. At that point "I didn't realize that he wasn't going to go. But I got the message pretty quick. He said, 'Hell, if I was gonna die, I'd already be dead!'" Accepting the firmness and logic of the old rancher's decision, Kennedy flew back to Brooks Field that night—alone.

Sometime later a runaway tractor almost succeeded where the rattlesnake had failed in terminating prematurely Elmo Johnson's Big Bend tenure. Again his frontier resiliency and hardheaded determination plus the Air Corps emergency medical service enabled him to survive and continue to reign over his riverside empire.

The tractor Johnson used to grade the runways required hand-cranking to start the engine. As a safety precaution, he always left the gear shift handle in neutral position. However, just prior to the tragedy, some children visiting the family had been playing on the tractor and accidentally forced the gears into high position. Failing to check the transmission setting before hand-cranking the vehicle, Johnson started the engine. The tractor leaped forward, striking him and pinning him beneath the rear wheel. "The left wheel hit him, [and] beat him down into a trench," Col. Don Mayhue recalled. "And the front end [of the tractor] hit something and couldn't move forward." So the unattended vehicle, engine running, wheels turning, and unable to move forward, "was standing there jumping up and down on him." Johnson finally freed himself from beneath the machine, and Mrs. Johnson and others who were at the trading post carried him into the house (interview, February 7, 1978).

Johnson's condition was obviously critical. While Mrs. Johnson attempted to summon local help, the radio operator stationed at the airfield dispatched an emergency message to the Eighth Corps Area headquarters describing the nature of the accident and requesting whatever help the Air Corps could provide. Mayhue, attached to the Twenty-second Observation Squadron at Brooks Field, received orders to fly the emergency mission. He alerted flight surgeon Capt. Edward Sigerfoos, and around midafternoon the two officers took off in a three-place North American O-47 for Johnson's Ranch.

They arrived late in the afternoon and found Johnson in bed, sitting up. His injury rendered a prone position unbearable. Sigerfoos' tentative diagnosis: a back fracture and possible additional internal injuries. There remained no question in the young medic's mind: hospitalization was mandatory. After giving Johnson an injection of morphine to ease his suffering, they began preparations for transferring him to the airplane. Johnson, however, had not been consulted in the matter. When it was suggested that he be flown to a San Antonio hospital, he opposed the move loudly and vociferously. "We had to argue plenty," Mayhue remembered, "to get him to even consider going to a hospital" (ibid.). Faced with further complications and possibly paralysis, Johnson was finally persuaded to go to San Antonio.

The O-47 was the Air Corps' latest observation aircraft, had a cruising speed of two hundred miles per hour, and carried a three-man crew seated in tandem beneath a sliding canopy. This particular design feature, plus its speed, accounts for Mayhue's selecting the O-47 for this mission. The pilot sat forward, the observer/photographer occupied the midsection, and the gunner sat in the rear seat, which they prepared for Johnson's trip to San Antonio.

Sigerfoos faced two problems with his patient. First, he had to remove Johnson, with a possible back fracture, from his bed to the airplane; and second, once there, considering the possibility of other undiagnosed injuries, he had to lift him prone into the rear seat through the narrow canopy opening. The young medic prepared his patient for the ordeal with a second injection of morphine and began the transfer. They lifted Johnson from the bed, placed him in a straight-backed chair, and carried him to the waiting airplane. "We got him up on the wing," Sigerfoos remembered, "and eased him in the middle seat. Somebody got in the back [seat], and with three people on each side, we got him in the airplane and got him comfortable" (interview, February 8, 1978).

To add to Sigerfoos' anxiety, it was getting late, and "I remember Don [Mayhue] was getting a little anxious because we weren't going to stop for gas on the way home. His airplane only had a 150-gallon fuel capacity." But once they had their patient comfortable, Mayhue taxied out to the runway and, just before sunset, took off and flew directly to Brooks Field. They landed after dark and taxied up to the ramp where Sigerfoos had an ambulance waiting. Removing Johnson from the O-47 was even more difficult than getting him in. "We had [an] air mattress," Sigerfoos recalled, "and we could maneuver it around and we finally got him out [and] put him in the ambulance" (ibid.).

Subsequent diagnosis revealed Johnson had indeed sustained compound fractures of the back and pelvis. He was placed in a cast and remained in the hospital about two weeks. When he was released, his daughter took him to her home in San Antonio, where he remained another two weeks, complaining daily about returning to the ranch.

Johnson's concern for his livestock, the airfield, and the trading post was too great to allow such matters as fractured bones and a steel brace to restrict his activities. "The last time I saw him," Mayhue recalled, "he was wearing a steel brace. But, in the meantime, he had been cutting pieces off [the brace] and riding a horse." Johnson, however, complained of persistent back pains, which he attributed to the annoying brace. Two West Texas doctors he consulted thought otherwise. They recommended that he return to San Antonio for further examinations, and if the fractures had healed sufficiently, have the brace removed by qualified personnel. "That made 'im mad," said Mayhue, recalling the outburst. "So, he went back down in the Big Bend, got out the meat saw, and had his wife cut that brace [or cast] off" (interview, February 7, 1978). And the Big Bend country returned to normal.

Most of the Air Corps pilots who flew to the riverside airfield for the weekend remained for Sunday lunch (the Johnsons called it dinner) and usually scheduled a midafternoon departure, allowing sufficient daylight for the return flight. Despite the area's sparse population, the row of Air Corps planes parked beside the runway never failed to attract a host of uninvited but not unwelcome Sunday afternoon spectators. The visitors were free to inspect the aircraft, and some were even given courtesy sight-seeing rides in the military planes, a practice strictly forbidden by Air Corps regulations.[2] The spectators, as well as the pilots, were acutely aware of this restriction; if a ranking officer was visiting the Johnsons, the matter of complimentary airplane rides went unmentioned. General Parker, then a second lieutenant, recalled aborting a take-off in his BT-2 with a civilian passenger when 1st Lt. Thad Foster landed unexpectedly at the ranch. Parker quickly ordered his surprised passenger out of the aircraft and taxied back to the ranch house. Foster asked no questions; Parker offered no explanations (interview, April 13, 1978).

Although Foster enforced the nonmilitary mandate among the junior officers, he obviously felt no compunction about sharing his aircraft with civilians, especially attractive young ladies. Lorene Green,

---

2. The War Department further reinforced this ruling in 1933, "issuing orders prohibiting the use of Army planes by non-military passengers, except in cases of emergency involving 'life or catastrophe.'" The justification: lack of funds and increasing demand by nonmilitary passengers (*Aero Digest* [November 1933]: 62).

Sunday afternoon visitors at Johnson's Ranch pose in the shade of Lt. "Lefty" Parker's Douglas BT-2. Parker's later attempt to give G. E. Babb (in white hat and pants) a complimentary airplane ride was interrupted by Lt. Thad V. Foster's unexpected arrival. Courtesy Mrs. G. E. Babb.

Elmo Johnson's niece, became a recipient of the young lieutenant's favors. She shared her uncle's affection for Foster, knowing

> he was quite a character, and had a very charming personality, also well mannered and very daring in his flying. One of the flights I made with him was over Santa Elena canyon. When we were getting ready to leave the canyon area, he flew right down between the cliffs on the east end of the canyon and then straight up. We then buzzed Castolon, then flew between Mule Ear Peaks and back to the ranch. He was very careful to see that you had a parachute and helmet on. Being rather young, it didn't frighten me to fly with him. I am sure he was one of the best aviators. (Correspondence, September 30, 1980)

Flying civilians on Sunday afternoon sightseeing trips was one thing; invading the airspace of a foreign country in pursuit of criminals while flying government aircraft was an entirely different matter. When Lts. Charles Deerwester and B. A. ("Bunk") Bridget arrived at Johnson's Ranch in two Consolidated PT-3 primary trainers from Brooks Field, they realized immediately that something was wrong. An unusually large number of armed men had gathered in front of the trading post. Elmo Johnson introduced them as lawmen investigating a recent double murder that occurred near the ranch. According to the reports, Pete and Carlos García, brothers-in-law, went to the home of Pablo Baisa to discuss a personal matter with him and several other men. The conversation became heated, and in the ensuing fight both young men were killed (*Alpine Avalanche*, February 7, 1930).

Lawmen from two nations converged at Johnson's Ranch to investigate the double murder. Mexican *rurales* subsequently located the trail of the fleeing murderers across from where it was believed the bodies were thrown in the Rio Grande. Following this trail some twenty miles into Mexico, they located an abandoned building in which the suspected murderers were believed to be hiding (ibid.).

Johnson had a personal interest in the matter. The two victims, not related, were married to two sisters, María and Andrea Holguín, both of whom worked for the Johnsons. Recognizing Johnson's personal concern, Deerwester and Bridget offered their services, including the use of the two United States Army Air Corps planes, in helping locate the bodies and bringing the murderers to justice. Operating from a military base established to protect the region from depredations south of the border, the United States airforce was again responding to the call for help, but this time in a manner highly unofficial and totally unauthorized. This meant invading Mexican airspace without the benefit of diplomatic negotiation or approval by the United States Department of State. But who was there to be consulted at Johnson's Ranch?

The first objective was to locate the bodies. Since Johnson knew the country well on both sides of the Rio Grande, he joined Deerwester in one PT-3 and they took off for Mexico. They flew south to the point where the murderers crossed the river and began their aerial search. Johnson recalled that he watched the ground for a trail left in the wake of the fleeing horsemen. He soon identified a vague pattern of hoof-prints, which they followed toward the interior of Mexico; however, the trail suddenly reversed its course, doubling back toward the Rio Grande. Responding to Johnson's hand signals, Deerwester turned the open cockpit biplane around and began following the trail northward. The fleeing murderers apparently planned their circuitous escape route to confuse a pursuing ground party. They had no way of anticipating Deerwester's and Johnson's aerial participation in the search.

As they approached the river, Johnson saw something suspicious below him and motioned to Deerwester to circle the area for a closer inspection. As they flew over an abandoned goat pen, Johnson saw what appeared to be a human leg protruding from freshly moved earth. Deerwester circled the site again; on closer inspection it became apparent what had occurred. The fleeing killers had dumped the bodies in a ravine, caved off the soft dirt banks to cover them, and then drove a herd of goats over the improvised graves to camouflage the burial site. Believing they had located the graves of the two victims, Deerwester and Johnson returned to the airfield to report their discovery to the waiting lawmen. An on-site inquest confirmed their findings. "The bodies were left where found on the other side of the river," the *Alpine Avalanche* reported, "as Mexican authorities would not permit their removal to this side" (February 28, 1930).

The fugitives, however, remained at large. While the Texas authorities secured the north side of the river, the *rurales* advanced toward the suspected hideout to identify and question the occupants. Dividing the two law enforcement groups posed a problem of communication; Deerwester again provided the solution in his PT-3. "I had a [Texas] ranger in the airplane," he recalled. "We were dropping messages to the *rurales* on the other side so they could catch those people. . . . Our rangers were backing them up so that, if they did come into the United States, they would have some support—either pick them up or chase them back across the river" (interview, February 11, 1971).

The suspected murderers, however, failed to appear at the border. In order for the Texas authorities to file charges against them, positive identification was necessary. Again both Air Corps pilots cooperated in this effort. "We made various flights, and I think everyone had a chance to get up and take a look," Deerwester explained, "because Bunk [Lieutenant B. A.] Bridget and I were both there" (ibid.). W. D. Smithers, who flew with Deerwester on one flight, described their "bombing run" of the fugitive's hideout as follows:

> They thought that these two Mexicans, Arocodia Billalba—we
> called him Jake—and one of the Baisas were hiding out in the
> house that sat up on top of the hill. The rangers figured they were
> in there, but weren't sure. They [the rangers] wanted to know if
> there was any way to get them out of the house to see if they were
> the ones they were looking for. So Deerwester said he could get
> 'em out with one of those smoke flares. So he flew down level with
> the house . . . pulled the tab [on one of the cannisters] and threw

that container right square through the door. . . . It had a stream of yellow smoke and those Mexicans really came outta there. They thought it was a bomb. It was those two Mexicans they were after, but they were in Mexico, you see. (Interview, August 27, 1975)

Deerwester's low-level attack on the hilltop hideout apparently yielded results. The week following the murder, the *Avalanche* reported a third arrest had been made in the case, and "a second son of Pablo Baisa is now in jail being held in connection with the slaying" (February 14, 1930).

Although Johnson expressed personal concern for the welfare of the Holguín sisters, widows of the murdered men, he attached slight importance to the entire matter. He later discussed the episode with Gen. William L. Kennedy, who recalled that Johnson "didn't think much about it. It was just part of life on the border" (interview, January 19, 1978).

Years later, reflecting on his unauthorized and illegal involvement in the matter, Deerwester was equally casual in his response. His explanation summarized completely the mutual bond that developed between Johnson and the more flamboyant members of the "Order of the White Scarf": "Elmo just asked us to help out and we volunteered" (interview, February 11, 1971).

"We didn't know it at the time, but we were learning how to fight one hellova war."

—Gen. William L. Kennedy

Scene 4.

Another War, Another World, and the End of an Era

"... I want airplanes—now—and lots of them"

CHAPTER 14

The ninety-eight landings recorded during 1934 and 1935 (July 14 to July 3) constituted the highwater mark of air traffic to Johnson's Ranch. The following year the field register entries dropped to seventy and continued to decline each year as the 1930s drew to a close.

During this period, political and military upheavals in areas far removed from the Big Bend country eventually came to bear on the flow of traffic to Johnson's Ranch. Italy's invasion of Ethiopia in 1935, the outbreak of the Spanish Civil War in 1936, the renewal of the Sino-Japanese conflict, and Germany's seizure of the Czechoslovakian Sudetenland in 1938 forecast the end of a tenuous era of peace. The runways of the Big Bend airfield would ultimately reflect the United States' responses to global instability.

Although isolationism dominated political thought throughout much of the decade, news dispatches from the foreign battle fronts conveyed a vastly different message to the Army Air Corps strategic planners. The message was irrefutable: air power held the key to victory in modern warfare. Developing an air force commensurate with changing world conditions became the challenge as the decade reached midpoint. But with limited funds, change came slowly and laboriously.

On July 28, 1935, the Boeing Airplane Company test-flew its latest product developed for the United States Army Air Corps—the B-17 prototype. During World War II, modified versions of this aircraft, the "Flying Fortress," became an American household word. By 1938, only twelve of these first-line, long-distance heavy bombers were in service. And the following year, when war began in Europe, the Air Corps inventory listed only nineteen (Glines, 1963: 135).

Although no B-17s appear on the Johnson's Ranch field register, other units of an expanding modernized air force began landing at the

Big Bend air station during the latter half of the decade. The aircraft entries on the final pages of the register included bomber, transport, attack, and multipurpose surveillance types, forecasting an impending world conflict in which many of Elmo Johnson's Air Corps colleagues would assume leadership roles.

The peacetime era of Air Corps history was gradually coming to an end. As the decade passed the halfway mark, however, officers still exercised a high degree of individual freedom. With time, fuel, and equipment at their disposal, improving pilot efficiency was still justification for an Air Corps pilot to employ military aircraft in a training/recreational operation.

Traditions, which the passage of time does not easily erase, grow slowly in the minds of the practitioners. The "Order of the White Scarf," the product of World War I romanticism, which grew with the Air Corps through the 1920s and into the 1930s, was destined to continue. But as the officers' numbers and responsibilities increased manyfold, their personal freedom diminished proportionately. Thus the aura of romanticism that once enveloped the "intrepid airman" gradually became diluted. Yet romantic myths exist within no well-defined lines of demarcation and its practitioners, committed to perpetuate a tradition, continued to forge a personal image that time and custom had imposed. But the drama was destined to end; the perpetual cycle was approaching full circle. A romantic tradition born of war would ultimately be lost in the tragic course of world events.

Johnson's Ranch, however, remained a stronghold for the Air Corps' free spirits. There was an occasional bandit to pursue, a river to strafe, antelopes to chase, canyons to challenge in open cockpit trainers. It was a place where everyone disregarded rank and Mrs. Johnson fed generals and privates at the same table, where Air Corps officers still "wore their hunting togs and carried their side arms and played sheriff—with Elmo bossing us all" (Deerwester, interview, Febuary 11, 1971).

Of all the free spirits that visited Johnson's Ranch, none exhibited the same casual disregard for military discipline as did Lt. Thad V. Foster. In spite of his carefree attitude, he nevertheless discharged his responsibilities with a high degree of professionalism. Maintaining the Eighth Corps Area auxiliary landing fields became Foster's mission in life. But of all the system's facilities, he regarded Johnson's Ranch with a marked partiality. In addition to helping establish the field, he maintained a regular inspection schedule, examining runway conditions, requisitioning fuel, supplying the transmitter unit, and renewing annually the Air Corps lease on the property.

Foster's friendship with the Johnsons accounts largely for his more-than-mandatory overnight stops at the Big Bend airfield. During the year ending July 3, 1935, he made a record fifteen inspection flights to the ranch. Changes in rank entered in the field register for this period indicate that his service record—his official record—had not gone unnoticed. Between July 6, 1929 and March 10, 1934, Foster registered as a lieutenant; however, on April 7, 1934, he entered his rank as captain. The additional gleam of his twin bars apparently gave Foster a renewed sense of security. Before the year ended, the names of three female companions appeared adjacent to Captain Foster's.

Mobility is a fact of military life. It was inevitable that the officer most closely identified with the Southwestern Airways facility would ultimately be relieved of his cherished duties. And while Foster may have accepted the change reluctantly, he recorded the event with flair and regional color, according himself additional rank many colleagues regarded as suspect. His June 4, 1935 entry, composed on two lines, remains the largest and most pretentious in the register:

T. V. Foster, Colonel, New Mexico National Guard
Last Inspection Trip—HASTA Luega [see you later]

On June 28, 1935, Foster registered again as a colonel in the New Mexico National Guard, but this time in the company of his replacement, Capt. Charles A. Pursley, air officer, Southwestern Airways. Foster made his last official flight to the airfield on July 3, 1935, again with Pursley, and listed his rank as captain based at Brooks Field.

Foster was absent from the Big Bend for more than a year, returning briefly on July 18, 1936, to visit the Johnsons. He retained his captain's rank, but in the meantime had been transferred to Maxwell Field, Alabama, site of the Air Corps Tactical School. Foster's register entry is uncharacteristically brief, containing only his name, rank, field, and aircraft. The latter is of some special significance. For this visit Foster chose a Boeing P-12 pursuit plane, the Air Corps' glamor symbol of the early 1930s, an appropriate vehicle for his farewell appearance at Johnson's Ranch.

Between Pursley's first inspection assignment on June 28, 1935, and his last on September 17, 1938, a period of slightly more than three years, he lists only twelve inspection flights in the field register. Lt. LeRoy (Joe) Hudson joined Pursley in 1936 in the administration of Southwestern Airways, and on August 8 they made an overnight trip to the ranch. No more inspection flights are recorded for the year. The following year the two officers each visited the field once on subsequent days. Hudson registered on June 16, 1937, and Pursley on June 17. These statistics reflect the Eighth Corps Area's declining interest in the Big Bend facility as well as the field's diminishing importance to the region's security. A new set of priorities preoccupied the Air Corps general staff.

From 1937 on, the field register reflects a premonition of things to come. By June 1937, annual aircraft landings had dropped to only twenty-one, the lowest traffic count since the first year of operation. The count for the next two years remained virtually unchanged. Twenty-two landings are recorded for the year ending June 25, 1938, with twenty-six landings entered for the same period in 1939.

By decade's end, assignment changes began eliminating many familiar signatures from the pages of the field register. Lt. William L. Kennedy registered the last time at Johnson's Ranch on November 20, 1937, and Lt. Hugh A. Parker, one of the Johnson's most devoted visitors, bid the family his final farewell on November 23, 1938. Lt. Don W. Mayhue, who had recently come to Elmo Johnson's rescue following the tractor accident, inscribed his final entry on August 10, 1939, and eleven days later, Lt. John W. Egan visited the Big Bend airfield for the last time. Many of these young Air Corps officers, who regarded Johnson's Ranch as a very special place, were now approaching command-level assignments. With their specialized skills and experience, the drive to expand the Air Corps' combat potential required their services elsewhere.

As air traffic to Johnson's Ranch declined, new aircraft types continued to appear on the field register. Some of the new equipment were commercial modifications; others were developed especially for combat assignments. All reflect the gradual buildup for the inevitable.

On May 27, 1937, Lt. Sigma A. Gilkey, attached to the First Transport Squadron at Patterson Field, Ohio, registered a Douglas C-33 at the Big Bend airfield. This was the Air Corps' cargo-transport version of the Douglas DC-2 commercial airliner, which at that time had been in airline service for more than three years. The C-33 was the immediate predecessor of the Douglas C-47 "Skytrain" (DC-3 commercial version), which evolved as the Air Corps' World War II aerial workhorse.

The following November 25, 1937, a flight of two Northrop A-17A attack planes piloted by Lts. J. D. Whitt and J. R. ("Killer") Kane arrived at the ranch on a flight from Barksdale Field, Louisiana.[1] These low-wing, all-metal, two-place attack planes were attached to the Ninetieth Attack Squadron of the Third Attack Group. With a maximum speed of 220 miles per hour and a range of 1,195 miles, the

---

1. J. R. ("Killer") Kane was one of the leaders of the low-level Army Air Force attack on Ploesti on August 1, 1943. Kane received the Medal of Honor for his role in this famed air battle.

Northrop A-17A had replaced the Curtiss A-12 as the Air Corps' first-line, low-level attack aircraft.

The pages of the Johnson's Ranch field register provide graphic documentation of the United States' step-by-step process toward becoming a major force in air power. The Northrop A-17A, for example, represents the third most advanced type of attack aircraft to land at Johnson's Ranch. From the Curtiss A-3 "Falcon" biplane, first registered on July 25, 1931, to the Curtiss A-12 "Shrike" entered on November 15, 1934, and finally to the Northrop A-17A, the Air Corps had, within approximately five years, achieved dramatic advances in its low-level attack capability. Speed, range, and firepower had almost doubled, while the lacquer-finished fabric stretched over a tubular steel framework had given way to an all-metal, skin-stressed construction.

The aircraft assigned to the Third Attack Group, however, remained in transition as the Air Corps procurement board searched for combat units possessing even greater efficiency and flexibility. For a brief period in the late 1930s, the Douglas B-18 became that group's first-line medium bomber. Lieutenant Kane returned to Johnson's Ranch on July 15, 1939, in a B-18, the largest aircraft ever to land at the Big Bend airfield. A modification of the Douglas DC-2 commercial transport, this low-wing, twin-engine, all-metal bomber carried a six-man crew and operated at a maximum speed of 215 miles per hour. The B-18 weighed 27,673 pounds with maximum combat load, more than twice that of the Martin B-12 bomber, which it replaced.[2] (By the time Kane landed the B-18 at Johnson's Ranch, the Third Attack Group had already taken delivery of the first of the Douglas A-20 "Havoc" low-level light bombers and attack aircraft. The A-20 never appeared at Johnson's Ranch; time and urgency dictated its deployment elsewhere.)

Equally significant changes were occurring in the Air Corps pilot training program, both in instructional procedures as well as in the new aircraft being developed for this purpose. As in the combat types, greater horsepower, higher performance capability, and more metal and less fabric characterized the new basic and advanced trainers. On September 16, 1938, Maj. Nathan F. Twining, then based at Duncan

---

2. When Lieutenant Kane flew the B-18 to Johnson's Ranch, its period of combat squadron assignments was nearing an end. A year later, however, B-18s and B-18As equipped most of the bomber squadrons, and the type was still in front-line service when Japan attacked Pearl Harbor on December 7, 1941. Many of these medium bombers, assigned to the Fifth and Eleventh Bomb Groups, were destroyed during that attack.

Field, chose a North American BC-1 (basic combat trainer) for his final flight to Johnson's Ranch. This series of two-place, low-wing, single-engine aircraft, developed by North American Aviation, Inc. (the AT-6 "Texan" or "Harvard" was the more famous version), formed the core of the Air Corps combat training fleet just before and during World War II.

Although accelerated flight training programs dominated Air Corps activities in the late 1930s, the Eighth Corps Area continued to maintain and operate the Big Bend radio station, but even that was destined to end. The last radio operator assigned to the post was Pfc. Kyle Johnson. He received orders in June 1938 to proceed from Biggs Field to Johnson's Ranch and relieve a Pfc. Williams for his annual thirty-day leave. Illness, however, prevented Williams' return and Johnson remained at the Big Bend air station about a year. (The following account of Johnson's transfer and closing of the radio station is taken from a letter from Johnson dated June 19, 1979.)

A Lieutenant Weston flew Johnson from Biggs Field to his new assignment in a Douglas O-38. Johnson occupied the cramped rear cockpit of the two-place biplane along with all of his gear, which included a footlocker, helmet, and rifle. After landing at the ranch, they taxied over to where the Johnsons and Williams were waiting. Weston parked the aircraft and Private Johnson began the awkward task of removing himself and his ill-fitting gear from the airplane. Johnson's ludicrous Big Bend arrival was equalled only by the gross incongruity of the exchange of protocol that ensued. "Back in those days," Johnson explained, "it was quite common and proper for a visiting pilot to report directly to the official in charge of a station." This official was, of course, Private Williams. So the young lieutenant, "sharp on protocol as well as flying . . . bowed to the Johnsons, turned to Williams, and with a beautiful West Point salute, said, 'Lt. Weston reporting as ordered, sir.'"

If Williams was amused by Weston's unexpected formality, he did not smile; he too was familiar with official protocol. But nothing seemed to fit—the gestures, the words, and above all the appearances. Being the sole member of a wilderness military establishment had instilled in Williams a casual indifference toward military routine as well as his personal appearance. Clad unceremoniously in shorts, a ragged T-shirt, and several days' growth of beard, Williams faced immaculately attired Weston, whose polished tan shoes and neatly pressed uniform radiated a personal reverence for military tradition.

The contrast between the two men standing at granite-like attention was startling and bordered on the absurd. Yet Williams joined in the farce and delivered his lines with convincing authority. Returning the lieutenant's salute, Williams responded, "'Welcome to Johnson's Ranch, what is your cargo?' Well, the lieutenant was not slowed a bit," Johnson explained. "He began naming the routine gear in my baggage (surprisingly accurate) and ended up with, 'and PFC Johnson who is your replacement!'" Recalling the event with obvious delight, Johnson added, "Old Mr. [Elmo] Johnson and I enjoyed this many times later when he retold the story to visitors."

If the new radio operator was dismayed by his farcical introduction to his new post, his first evening with the Johnsons demonstrated that the new assignment included some additional benefits.

We cleaned up—swam nude in the river—and went to dinner with the Johnsons. We had a fine meal of goat meat, tacos, and beans, beautiful tomatoes, lettuce and green onions from the garden. After dinner, Mr. J[ohnson] and the pilot, who stayed overnight, of course, went to the patio to test some old whiskey and talk about the outside world (San Antonio, Austin, Dallas, etc.) and Mrs. J[ohnson], Williams, and I went to the outer yard to where a picnic table, benches, etc. were located, to test some wine smuggled in from Old Mexico.

Private Johnson soon adjusted to the new military lifestyle that seemed compatible with the assignment. Casual apparel replaced his military uniform, and when not operating the radio transmitter he fished in the Rio Grande, gathered vegetables, and helped with the chores. He soon became part of the Johnsons' vast military family. And if Williams' failure to return to the Big Bend gave Johnson cause for concern, it never appears in his correspondence. But the assignment was destined to end; change was in the offing. Ongoing developments would alter the course of world events and bring to an end a romantic lifestyle that existed along the banks of the Rio Grande.

On September 28, 1938, British prime minister Neville Chamberlain returned to London from the Munich conference to naively announce to the world that there would be "peace in our time." German chancellor Adolf Hitler dictated the price of "peace"—Czechoslovakia. Unlike his adversaries, Hitler negotiated from a position of military strength; his brazen demands sent shock waves throughout the free world. In assessing the tragic consequences of the Munich appeasement, military analyst George Fielding Elliot termed it "blackmail made possible only by the existence of air power" (Glines, 1963: 46). His statement carried prophetic overtones.

Following Chamberlain's London announcement, President Franklin D. Roosevelt summoned his military advisors to the White House to discuss air power. Gens. George C. Marshall, Henry A. Arnold, and Marlin C. Craig, Secretary of the Navy Charles Edison, and Secretary of War Harry H. Woodring attended the conference. According to aviation historian Carroll V. Glines, the president "came right to the point: 'I want airplanes—now—and lots of them.'" He foresaw an additional aerial striking force of "10,000 fighting planes of all descriptions in production by 1940 and 20,000 the next year." He assessed the cost of his emergency proposal as "about $300 billion, including the cost of airfield construction and training personnel" (ibid., pp. 146–147).

In the wake of the president's demand for air power, pilot training received an immediate priority. Prior to 1939, only three hundred pilots had been trained annually; however, "the pilot training objective grew to an annual rate of 1,200 by the end of 1939, to 7,000 a year by June 1940, and a month later was changed to a rate of 12,000 annually" (ibid., pp. 150–151). The training system had to be expanded immediately and much of this accelerated activity centered on the established San Antonio–area bases—Randolph, Kelly, and Brooks fields. The onrush of cadet recruits, however, created a demand for additional instructors that opened new opportunities for experienced Air Corps officers. Many former members of Elmo Johnson's "Big Bend Escadrille" were summarily assigned training responsibilities.

A new concept of military preparedness was fast taking shape. Gen. William L. Kennedy explained how traditional methods and procedures no longer applied.

> It wasn't so much a change of policy as it was a change in mission. . . . The situation at Kelly [Field] was utterly different. Earlier, we had the class system and we had a personal relationship with each student. We flew with each student in accordance with his personality and tried to teach him to be the best goddamned pilot that ever came down the road. But once the expansion started, we had to do it on a production line basis. (Interview, February 20, 1979)

Administrative demands on the expanding training operation quickly elevated junior officers to command positions. Kennedy complained that he and John Egan "were moving up much faster than we wanted to." According to Egan, "We got to be captains by January 1939, and by the middle of 1942 we were full colonels." With rank came responsibility, and demands on the young officers' time increased proportionately. "We were flying day and night, seven days a week," Kennedy explained. Recreation became a luxury that no one could any longer afford. And for Kennedy and Egan, cross-country flights to Johnson's Ranch were just memories (ibid.).

As the pilot training program expanded in accordance with the War Department's upward manpower projections, the training command began searching for sites for new flight training centers. The airfield at Johnson's Ranch was considered and subsequently rejected. The reasons were obvious: it was located off the east-west traffic route and had poor ground communications, poor access roads, and no rail line.

Johnson's Ranch, like countless other auxiliary facilities maintained by the Airways Section during an earlier air age, was no longer needed. Johnson's Ranch, which almost a decade earlier had been the beneficiary of the horse cavalry's loss to mechanization, was itself now the victim of technological progress. By the late 1930s, military planes had longer ranges and the auxiliary fields once required for servicing stops were superfluous.

There remained no place for Johnson's Ranch in the Air Corps' immediate scheme of things. Maintaining an "isolated window into the happenings along the Mexican border" no longer interested army intelligence. The Escobar Syndrome had, at least temporarily, run its course. Landings dropped to twenty-two during 1939, with the field register entries showing a predominance of heavier, long-range combat types. Significantly, the traffic for the entire year was reported on only ten separate days: one day each in April and May, two days in July, three in August, one in September, and two in October.

The gradual disappearance of military aircraft from the long dirt runways that bisected Elmo Johnson's ranchland forecast the end of an era in Big Bend history. Sometime in early June 1939, radio operator Kyle Johnson received orders from Eighth Corps Area headquarters to close the station and move the equipment to Fort Sam Houston. A few days later two army trucks arrived at the airfield and the transfer began. Johnson recalled his mixed feelings as he bid the Johnsons farewell and embarked with the convoy. He had volunteered for a thirty-day relief assignment that lengthened into twelve months and left him stranded at the perimeter of civilization. Yet duty at the Big Bend airfield had netted him many personal benefits probably not found at any other Air Corps station: total freedom from military regimentation, an abundance of good food and good friends, plus time and accommodations to hunt, fish, and enjoy the pristine wilderness of the Big Bend country. "The Johnsons were good people and nice to me. I enjoyed my year there" (correspondence, June 19, 1979).

Spring gave way to summer, and with the approach of autumn came the harbinger of fate that military observers had long anticipated. In the predawn of September 1, 1939, Adolf Hitler unleashed his tanks and dive bombers on Poland, setting the stage for the beginning of World War II. With war in Europe a frightening reality, the Army Air Corps further escalated its flight training goals, expanded its facilities, and reassigned most experienced personnel in the rush to prepare for another war that now appeared inevitable. Air traffic to Johnson's Ranch, however, continued intermittently until October 21, when Lt. J. R. Kane, one of the Third Attack Group's more avid hunters, arrived from Barksdale Field in a Northrop A-17A. That was the last aircraft to use the Big Bend airfield that year. Not until the following July 28, 1940, does another entry appear in the field register. That, plus an entry dated September 16, 1940—only two flights—constituted the year's activity.

Days lengthened into weeks, weeks into months, and no airplanes came to Johnson's Ranch. Autumn rain collected in the holes and depressions of the abandoned runways, and with the coming of spring, blades of grass appeared in the clean earth where landing aircraft once etched recurring patterns of military activity along the Rio Grande. Gone too was the roar of aircraft engines echoing off the canyon walls, silent prophecy of a vastly different world that lay somewhere beyond the threshold of an unknown tomorrow.

". . . the airplanes would return to Johnson's Ranch"

CHAPTER 15

The Big Bend country of Texas occupies the heart of an immense unprotected borderland that stretches almost two thousand miles from Brownsville, Texas, to San Diego, California. Canyons, mountain ranges, deserts, and unpopulated wastelands, broken here and there by oases of farms, ranches, and towns, follow this international boundary. It is marked sometimes by a river, sometimes by intermittent boundary markers, maybe a fence or an unpaved road, and sometimes by nothing more than arithmetic bearings on cartographic charts that few have seen and still fewer can comprehend. This immense underbelly of the American nation lay vulnerable to foreign interests unfriendly to United States security; its length and remoteness offered easy access to a nation busily gearing up for what appeared to be a global encounter, its adversaries as yet unidentified.

As 1941 drew to a close, pieces of the global jigsaw puzzle began to fall into place. In Europe, Hitler had devastated London and was threatening to invade the British homeland. Meanwhile, in Washington, high-level negotiations about another potentially volatile area appeared to be floundering. Japanese ambassador Kichisaburo Nomura and Special Envoy Saburo Kurusu, meeting with Secretary of State Cordell Hull, stood unyielding on their demands as their country continued to deploy troops into Southeast Asia. Dispatches from Tokyo, dated December 4, 1941, seemed to preclude further negotiations, claiming the United States position in the matter was unacceptable. Two days later, on December 6, in a last-ditch effort to preserve peace in the Pacific, President Franklin D. Roosevelt addressed a personal message to the emperor of Japan. A State Department spokesman, quoted in the *El Paso Times* December 7 edition, termed the president's effort an "appeal to prevent a Far Eastern Explosion" (December

7, 1941). Roosevelt toiled in vain. By the time many of the subscribers opened their Sunday edition of the *Times*, the explosion that changed the world was already history.

News of Japan's "dastardly attack" on Pearl Harbor alerted the American nation to its immediate vulnerability. In a time of crisis, the military's first objective is maintaining the national security by establishing control and surveillance along the nation's perimeters. And the land perimeter most vulnerable to trespass bisected the continent from Texas to California.

For the nation as a whole, the Pearl Harbor attack came as a total surprise. Military leaders, however, aware of the deteriorating conditions in the Far East, had initiated training procedures to prepare for a national emergency.[1] Their main objective was to protect public utilities from possible sabotage. By midafternoon on December 7, units of the Texas Defense Guard, "with live ammunition in their rifles and fixed bayonets . . . took up stations to guard vital railroad bridges in El Paso and on one of the nation's main rail arteries [the Southern Pacific and Texas Pacific] and the El Paso Electric Company's power plant in the Upper Valley" (*El Paso Times*, December 8, 1941). The following day the First Cavalry Division, stationed at Fort Bliss, relieved the Guard units and began patrolling the United States border east and west of El Paso "in cooperation with the friendly Mexican Army." The official announcement stated further that "planes of the 120th Observation Squadron of Biggs Field joined in the patrol" (ibid., December 9, 1941).

In view of the Fifth Column infiltration of Oahu prior to the December 7 attack and recent reports of increased Japanese activity in Mexico, the United States military was leaving nothing to chance. Mexican president Manuel Avila Camacho was equally apprehensive, initiating swift action against bands of armed Japanese fishing boats reportedly active in Baja California.

Secret and confidential G-2 intelligence reports, issued almost daily by the Southern Defense Command (SDC) headquarters, confirmed newspaper reports that colonies of foreign nationals were also residing near the United States border. Cited variously as "enemy combat operations," "enemy capabilities," and "other enemy activities," Report

---

1. "On 17 October 1941, for the purpose of implementing Southern Defense Command . . . the Commanding General, VIII Army Corps, was designated the Commander of the Southern Land Frontier, Southern Defense Command, and charged with the preparation and execution of the necessary plans for the defense of the frontier against external attack" (*Historical Record, Southern Defense Command, Phase I*, pp. 5–6, RG 338).

No. 4, dated January 2, 1942, stated: "Colony of Japanese farmers re-
ported about 50 miles south of Juárez, Mexico, opposite El Paso,
Texas, four of whom have adjacent fields suitable for landing fields"
(RG 338). The Japanese were subsequently removed to the interior of
Mexico.

The national hysteria that followed Pearl Harbor manifested itself
in many forms. Invasion rumors were rampant. On February 27, 1942,
Maj. Gen. I. P. Swift, commanding general of the First Cavalry Division, received an anonymous letter stating that the "Japanese will
feint a Pacific Coast attack to draw attention and will then bomb San
Antonio, and El Paso, Texas, via Mexico" (ibid.).*

Much of the intelligence data reaching the Southern Defense Command headquarters concerned unidentified aircraft sightings along the
United States–Mexican border. Considering the hundreds of miles of
this mostly remote, uninhabited, and largely unprotected frontier,
these sightings could not be ignored. Some of the official G-2 reports
are as follows:

Unidentified plane observed from Devil's River, Texas, 13 miles
northeast of Del Rio, Texas, flying east at 12:25 PM, 4 Jan. (January 8, 1942)

It is reliably reported that flares were observed dropped from
an airplane at the following points:
(1) Ten (10) miles west of Alpine, Texas, on the night of 7th
of January.
(2) In the vicinity of mouth of San Francisco Creek eighteen
miles (18) south of Sanderson, Texas, on the 7th and 8th of
January . . .
It is reliably reported that an unmarked, unidentified tri-
motored silver airplane was observed over Eagle Pass, Texas, at
1:30 PM, 3:30 PM and 7:45 PM, 9th of January. The same plane was
over hydro-electric plant near Eagle Pass, Texas, at 7:50 PM, 9th of
January. (January 11, 1942)

Same aircraft type reported southeast of Devil's River bridge,
3:30 PM, 11 January. (January 15, 1942)

Outposts of 112th Cavalry fired on unidentified plane flying
northeast over Pecos High Bridge. (February 17, 1942)

Four (4) large unidentified planes observed flying E. in Mexico high over Castelon [sic] at 10:20 AM, 2 March, then crossed to
U.S. and headed NE. (March 2, 1942) (Ibid.)

The G-2 reports contain no information if these sightings were ever verified as friendly or enemy aircraft. One fact, however, remains abundantly clear: unprecedented aerial activity was occurring along the United States–Mexican border. And with the ever-present fear of invasion gripping the Southwest Borderlands, every report, every rumor, however trivial, required investigation.

False emergency reports frequently sent the planes of the 120th Observation Squadron into action. Col. Henry L. Phillips, a former squadron member, recalled a late-night scramble to investigate a fire reported on Franklin Mountain in the shape of an arrow pointing toward Fort Bliss. Phillips and the squadron commander "rushed out to the field and took off. We flew out there and couldn't find a damn thing." Another emergency he investigated was somewhat more productive. This time Phillips discovered that the strips of freshly turned earth north of the base, also in the shape of an arrow pointing toward Fort Bliss, were nothing more than excavated laterals of a septic tank drainage system (interview, January 29, 1980).

In order to monitor all possible invasion routes via both land and air, the government quickly restricted many civilian aviation activities. The immediate objective: clear the skies of all nonessential air traffic. The same day the United States declared war on Japan, the Civil Aviation Authority announced the temporary suspension of all private aircraft pilot certificates except on scheduled airlines. The military also imposed additional restrictions on private flying, in some cases even forbidding civilian spectators at municipal airports. "Civilians are not allowed to visit the airport as they have in the past," announced the Fort D. A. Russell post commander. And, in addition, "no plane of any type, except Army planes, will be permitted to leave the airport, Marfa, Texas, without written permission of the commanding officer" (*Big Bend Sentinel*, December 26, 1941).

The region of Texas most vulnerable to infiltration through Mexico remained the Big Bend country. Cognizant of the region's history of border violations during periods of political stress, the United States Army quickly dispatched troops to that area. Dr. Ross A. Maxwell, director designate of the proposed Big Bend National Park, visited the area in December 1941 and witnessed the Big Bend military buildup. Wherever he traveled he encountered detachments of armed soldiers. Almost overnight Lajitas, Terlingua, Castolon, Johnson's Ranch, Glenn Spring, Hot Springs, Boquillas, and Persimmon Gap became armed villages. In addition to the soldiers, Maxwell explained that

the United States Geological Survey put in a string of bench marks [permanent markers indicating the location in terms of longitude and latitude] from Marathon on down to the river along roads that could be traveled at that time. One every mile so that they [the military] could get their exact location. In case of an invasion, they would know where it [the action] was according to a certain number on a bench mark. (Interview, November 29, 1978)

Late in December the Texas Rangers joined in the effort to seal off the southern boundary from alien entry. Capt. Ernest Best, commander of Company "E," stationed at San Angelo, led a task force of Rangers, four customs officers, and two border patrolmen "in a survey to prevent European aliens from entering this country across the Mexican border." Their findings were highly classified; "no information concerning this survey was divulged." (*Alpine Avalanche*, December 26, 1941).

By early 1942, when the feared invasion had not materialized, the army began withdrawing the border units for service in more critical combat areas. With Mexico's declaration of war on June 2, 1942, and the gradual deterioration of enemy capability on the continent, the need to maintain large numbers of troops along the Mexican border had lessened. The military, however, still considered the region vulnerable to alien penetration and began assigning other units to fill the gaps left vacant by the departing ground forces. History again seemed to be repeating itself; the unmistakable sounds of air power dramatized the changing of the border gaurd. The clatter of jeeps and the rumble of personnel carriers gradually gave way to the resonant drone of airplane engines. Older borderlands citizens who recalled the post–World War I era of border strife and later the brief reign of General Escobar knew the sound well. A new generation of combat aircraft was watching over the region, bringing with it a sense of security to those living along the Rio Grande. The airplanes were returning to Johnson's Ranch.

Immediately following the declaration of war, orders were issued to the 111th Observation Squadron (minus one flight) stationed at Fort Clark, Texas, and the 120th Observation Squadron based at Biggs Field, El Paso, Texas, "to fly not less than two reconnaissance missions during the daylight at irregular intervals covering the International Boundary of the Southern Land Frontier. All of these dispositions were completed by 11 December 1941."[2] The two squadrons divided

---

2. *Historical Record, Southern Defense Command, Phase II*, p. 5, RG 338. The 111th Observation Squadron, formerly a Texas National Guard Unit, was inducted into federal service on November 25, 1940, at Houston, Texas. After December 7, 1941, the 111th was divided into two echelons. The unit based at McAllen, Texas, flew antisubmarine missions along the Texas Gulf Coast. The 120th Observation Squadron, originally a Colorado National Guard unit, was ordered to active duty in the autumn of 1941.

the patrol assignment. Col. Henry L. Phillips recalled the division point was San Francisco Creek, which enters the Rio Grande at the Brewster County–Terrell County line. The 120th Squadron patrolled from that point westward to the New Mexico–Arizona border, while the area between San Francisco Creek and Brownsville, Texas, became the responsibility of the 111th Squadron (interview, January 29, 1980).

With Johnson's Ranch still a designated auxiliary field, pilots from both squadrons began landing their North American O-47s there shortly after the declaration of war. Facilities were limited to drums of emergency aviation gasoline and the Johnson's traditional hospitality. Five undated entries from Fort Bliss and Biggs Field began a new section of the Johnson's Ranch field register. Of the subsequent fourteen entries, with dates ranging from January 12 through February 20, 1942, four are from the 111th Squadron, with the remainder from the 120th.

One entry remains graphic in Phillips' recollection. After graduation from the Observation Flying School at Brooks Field, he joined the 120th Squadron in October 1941 and had about three months of peacetime duty before Pearl Harbor. Frustration and confusion accompanied the shift to wartime responsibilities, and "nobody ever told us exactly what we were supposed to do on this border patrol. We never did get any precise orders," Phillips explained. "If we saw somebody, were we to force 'em down, were we to shoot 'em down, or if they thumbed their noses at us, what action we should take? No one ever told us" (ibid.).

Phillips' unscheduled visit with the Johnsons occurred in late January 1942 while on a patrol mission between Biggs Field and Fort Clark. The crew in the three-place North American O-47 consisted of Lt. William H. Trachsel, the pilot; Captain Phillips, the observer-navigator; and Pfc. Robert N. Horn, the rear gunner. On the return segment of the flight, they planned to land at Marfa for fuel. Shortly after leaving Fort Clark they encountered unexpected bad weather and, after climbing to about 16,000 feet, were still in the overcast. With fuel running low, Phillips began calling military bases on the radio in Morse code requesting directions to alternate landing fields. Learning that all fields within their fuel range were "completely socked in," and unable to alert Biggs Field to his predicament, Phillips and Lieutenant Trachsel agreed that Johnson's Ranch was the only alternate field they could hope to reach. (With the radio station closed, they could gather no weather data on that facility.)

Descending from 16,000 feet, Trachsel was able to reduce power and conserve fuel. Shortly they began noticing thinning spots in the overcast. A few minutes later they "broke contact" and Phillips recognized a familiar section of the upper Big Bend country. He altered their course toward Johnson's Ranch and some twenty minutes later they were safely on the ground. Trachsel checked the fuel after landing and discovered that less than ten gallons remained in the reserve tank. Reflecting on his good fortune (and good navigation, which he did not mention), Phillips explained, "Marfa was closed in, and if Johnson's Ranch had also been closed in, we would have had to get out and use Mr. Irvin's safety descender [his parachute]!" (ibid.).

It was late in the afternoon when Phillips landed at Johnson's Ranch. With no way of determining the weather conditions at Biggs Field, he elected to remain there overnight. Other than the absence of the radio station, nothing had changed at the Big Bend air station; Phillips and his crew received the same hospitality as that accorded their colleagues from the prewar Air Corps. Mrs. Johnson and a maid prepared dinner for their unexpected guests. They slept in the main house and arose the following morning to an equally appetizing breakfast. Recalling the generosity of his hosts many years later, Phillips explained that the accommodations were gratis, as the Air Corps provided no per diem while on patrol. And the Johnsons neither asked for nor expected any compensation.

Shortly after Phillips' unscheduled visit to Johnson's Ranch, the 111th Observation Squadron was reassembled and moved to Augusta, Georgia. Leaving behind the North American O-47s, the squadron began training on Douglas twin-engine A-20 attack aircraft and Republic P-43 single engine fighters. This new equipment signaled a new age of reconnaissance aviation. On November 9, 1942, the squadron embarked from West Palm Beach, Florida, on a five-stop aerial ferry assignment to the Gold Coast of Africa. The previous day the Allies had launched the North African campaign.

With the departure of the 111th Observation Squadron from Texas, the 120th assumed the responsibility for patrolling the entire border from Brownsville, Texas, to the New Mexico–Arizona line. Biggs Field remained the central base of operations, dispatching twice-daily flights in both directions. This phase of the patrol operations, however, was brief; the 120th Squadron was also reassigned shortly thereafter.[3] Before the transfer, Phillips left the squadron to attend the

---

3. The 120th Observation Squadron, originally attached to the Eighth Army Corps on October 28, 1941, "was purely a training unit and no effective use could be made of it without interfering seriously with training activities. Accordingly it was released from this attachment shortly prior to 14 July 1942" (*Historical Record, Southern Defense Command, Phase II*, p. 33, RG 338). The 120th Observation Squadron was later disbanded at Birmingham, Alabama, on November 30, 1943, and the personnel reassigned.

Command General Staff College at Fort Leavenworth, Kansas. In retrospect, he regards this transfer as a good omen; the mission of observation was approaching a tactical milestone.

As the 111th and 120th Observation Squadrons patrolled the United States–Mexican border during the early days of World War II, traditional observation aviation was entering the twilight of military service. Experience proved that the peacetime doctrine of aerial surveillance and reconnaissance was totally incompatible with the realities of modern warfare. "I don't think there were many survivors from the 111th," Phillips recalled. "They were shot up completely." Combat statistics support his recollections; twenty-five officers of the 111th Observation Squadron were killed in combat. Of the twenty-five fatalities, twenty-one occurred in 1943 and 1944 in the North Africa and Italian campaigns.[4]

The open-cockpit, biplane observation aircraft that helped bring security to the Big Bend country—the O-2, the O-19, the O-38, and even the more advanced O-47—were ineffective descendants of a World War I Neanderthal, the De Havilland DH-4. Although this wood, wire, and fabric contraption helped switch the military spotlight from the ground to the air, its immediate offspring (products of an Air Corps wedded to the army) evolved as an earthbound concept, naively ignoring the enemy's retaliatory potential.

The Thomas-Morse O-19 in which Lt. Joseph H. Hicks searched for Mexican bandits along the Rio Grande in 1933 as well as the North American O-47 that Lt. William H. Trachsel landed unexpectedly at Johnson's Ranch almost a decade later are simply benchmarks in the total scope of combat aviation. Historically, these transitional development models remain an essential step toward technological progress, but they inevitably became useless in a war far removed from the era of the DH-4 as well as Johnson's Ranch.

---

4. For a comprehensive study of the changing role of observation aviation during World War II, see Futrell, 1956.

". . . they landed on sandbars in the Rio Grande"

CHAPTER 16

Following the transfer of the 120th Observation Squadron, United States intelligence authorities set in motion alternative plans to seal the cracks in the country's southern defense perimeter. With the armed forces committed to a two-hemisphere war, military aircraft could no longer be spared for this assignment.[1] It was the military, however, that turned to the last remaining manpower source, the civilian aviation sector, to maintain surveillance along the still critical United States–Mexican border.

Gen. Walter Krueger, commander of the Southern Defense Command, under whose authority the 120th Observation Squadron had operated, initiated the proposal. "Believing that surveillance of the Mexican Border from the air would be of value," he issued a request on July 24, 1942, "that Civil Air Patrol planes be provided the Southern Defense Command for that purpose."[2] Col. Harry H. Blee, Civil Air Patrol operations officer, assigned coordinating officer Maj. Harry K. Coffey of Portland, Oregon, the responsibility for establishing the new patrol. This veteran civilian pilot—he first soloed in 1914—

---

1. The Texas-Mexican border was not left unprotected. Ground troops were deployed throughout Texas and a troop of cavalry was stationed at Columbus, New Mexico. Motor and horse units patrolled the border at least three times a week.

2. *Historical Record, Southern Defense Command*, p. 33, RG 338. On December 1, 1941, Fiorello H. La Guardia, director of the United States Office of Civilian Defense, signed an order under presidential executive authority creating the Civil Air Patrol. D. Harold Byrd, Dallas oil executive who was instrumental in forming the CAP, was appointed commander of the Texas Wing.

embarked in late August 1942 for Texas to develop plans for the new operation.

Major Coffey's borderlands mission began at the Southern Defense Command headquarters in San Antonio, Texas, where he met with Col. William C. Crane, deputy chief of staff, and Col. Carleton F. Bond, air officer. This marked the first of several cordial, informative, and supportive conferences Coffey attended along the projected patrol route. The two officers shared all pertinent files and records with Coffey and explored the logistics of establishing the new patrol. They recommended further that Coffey conduct an aerial survey over the patrol route and arranged a conference with Gen. Harry Johnson, commanding general of the Southern Land Frontier, for the afternoon of August 27, 1942. At Fort McIntosh Coffey again found a receptive audience; General Johnson cited Coffey's proposal as "the first concrete evidence of constructive thinking and the first to have any plan of observation."[3]

The report of Major Coffey's mission reflects the thoroughness with which he approached his assignment. During his nine-day, 1,048-mile survey flight from the mouth of the Rio Grande to Douglas, Arizona, he identified emergency landing areas (including Johnson's Ranch) and canvassed the region for cooperating agencies and individuals. The federal counterespionage authorities Coffee met—United States Immigration and Naturalization Service, FBI, army and navy intelligence—confronted him with challenging evidence: ground surveillance alone could not secure the United States–Mexican border against foreign penetration. He found the most convincing evidence at Fort Bliss. Reports of sixty-one unidentified aircraft making border crossings convinced Coffey that these "detailed reports by seasoned observers . . . warrant the concern of the War Department, as this 1,050 miles of border is really vulnerable for the importation of undesirable persons and espionage agents" (CAP Records).

The War Department apparently concurred with Major Coffey's assessment. He completed his survey mission on September 4, 1942, and on September 10 submitted a comprehensive plan of operation to General Krueger. The general, in turn, recommended to the commanding general of the Army Air Force that Coffey's proposal "be approved and put into effect without delay." He concluded, "It is desired that the service . . . be designated 'Southern Frontier Liaison Patrol'

---

3. Major Coffey's Report, August 27, 1942, Civil Air Patrol Records, Col. L. E. Hopper, Personal Collection. All materials from this collection cited hereafter as CAP Records.

rather than 'Mexican Border Patrol.'"[4]

With the final approval of Coffey's plan, the Civil Air Patrol headquarters began negotiations with local military commanders for bases and sub-bases out of which the patrol could operate. Coffey established two administrative command centers, with Patrol No. 1 based at the Laredo Flexible Gunnery School, Laredo, Texas, and Patrol No. 2 based at Biggs Field, El Paso, Texas. He further cited Del Rio, Texas, a sub-base and Marfa, Texas, a reconnaissance base and recommended that refueling facilities be established at Brownsville, Texas, and Douglas, Arizona, for aircraft serving those areas.[5]

The safety of the flight personnel, both in the air and on the ground, was paramount in Major Coffey's planning. He recommended side arms for all pilots, observers, staff, and command personnel; two Thompson submachine guns in each ship in flight; rifles for all bases for all personnel except women; parachutes for flight personnel between El Paso and Del Rio; plus canteens, pocket compass, first-aid kit, U.S. Army emergency rations, and field glasses in each patrol aircraft. Coffey considered the Big Bend–Santiago Mountains area the most hazardous segment of Patrol No. 2 and cited eleven emergency landing fields in that region, which included Johnson's Ranch. At four of these facilities—Marfa, Presidio, Dryden, and Johnson's Ranch— he recommended "the organization of ground crews . . . and the use of two aircraft off each of . . . the fields to do reconnaissance and patrol work" (CAP Records).

With all the operational framework approved, all Major Coffey needed was pilots, airplanes, and ground support personnel. Most of these would come from the West Coast, where civilian flying had been restricted along a 150-mile Pacific Coast defense corridor.[6] Coffey ap-

---

4. This designation was subsequently altered. On December 7, 1942, Major Coffey telegraphed Col. Harry H. Blee: "We assured General Harry Johnson today the name CAP Southern Frontier Liaison Patrol will be changed to CAP Southern Liaison Patrol. The word Frontier seems to have offended our Mexican allies." Colonel Blee had previously ordered the change effective November 10, 1942 (CAP Records).

5. Early in 1943, Patrol No. 1 was removed from the gunnery school to Fort McIntosh, Texas, because the Army Air Force required the facilities. On December 1, 1943, Patrol No. 2 moved from Biggs Field to Municipal Airport, El Paso, Texas, for the same reason (*Historical Record, Southern Defense Command*, p. 34, RG 338).

6. With military, civilian, and commercial aircraft all flying in the congested Pacific Coast defense corridor, the problem of identifying enemy aircraft was increased substantially. Hence the West Coast restriction on civilian

pointed Maj. Jack R. Moore, a Portland, Oregon, wholesale distributor, to be commander of the Laredo base, and he, in turn, recruited most of his staff from that state.

Major Coffey wrote Maj. Bertrand Rhine, California CAP wing commander, on September 23, 1942, requesting that group supply the personnel and equipment for the El Paso base. This encompassed a total of 101 personnel and twenty-three aircraft, plus three ground-based and one portable radio station. There was urgency in Coffey's request: "We are now ready to start operation and request you expedite your enlistments of all personnel" (ibid.). Rhine responded by appointing Pasadena Secret Service agent Maj. Charles B. Rich as the El Paso base commander. His staff became known locally as the "California group."

At the outset Moore's and Rich's command assignments were vacant titles. When Rich reported to Major Coffey in El Paso on October 19, 1942, he found only one serviceable aircraft, "very little finance, no credit, and [we] were totally unknown to both the Army and Civilian population of El Paso." Moore was somewhat more fortunate; he brought with him a minimum staff whose personal property filled the void left by the Office of Civilian Defense. Moore's operation officer, Lt. William E. Lees, Jr., of Ontario, Oregon, called upon his recruits "to furnish all useful equipment available, such as mobile trailer houses, station wagons, mobile power units, radio transmitter and receivers. As a result . . . this unit went into operation with the nucleus of all the required equipment" (Moore and Rich, personal narratives, ibid.).

In spite of the individual commitments made by the patrol volunteers, Major Rich explained that

> the whole operation was handicapped from the beginning because we were required to get priorities through the War Production Board who had never been advised of the importance of our work and saw no reason why private airplanes should be given priorities. . . . The Army Air Force was likewise handicapped because their orders were not clear enough to allow us to either purchase or draw on receipt our much needed equipment. (Ibid.)

---

flying during World War II. Gen. John L. DeWitt, commander of the Western Defense Command, made no exceptions in the matter. According to Bertrand Rhine, California CAP wing commander, DeWitt stated he was "going to keep all civilian airplanes out of the air for the duration. Apparently DeWitt is not even happy about the airlines flying on schedule" (Bertrand Rhine to Earle J. Johnson, April 12, 1942, CAP Records).

Bureaucracy on top of bureaucracy bred more bureaucracy, and while the Office of Civilian Defense apparently understood the importance of the patrol operation, its allocations, in terms of civilian staff and equipment readily available, were, at the outset, visionary. Patrol No. 1 at Laredo was officially allocated sixty-one personnel, including ten pilots and ten observers and thirteen aircraft. Ten of these aircraft were to have less than ninety horsepower; three were to have 145 horsepower or more. El Paso Patrol No. 2 was to be the larger facility. Major Rich's authorized staff consisted of 102 specialists, including nineteen pilots and nineteen observers operating twenty-four aircraft. (These figures differ slightly from Major Coffey's original request.) Of these, eight would have less than ninety horsepower, eight 225 horsepower or more, and eight specified as "miscellaneous for Reconnaissance." Even in the face of acute staff shortages, the CAP still gave women a minimal role in the border patrol and restricted them to landbased duties. Colonel Blee ordered that "in no case will women be used as Pilots or Observers or assigned to any positions with the ground element other than those herein specifically authorized."[7]

In launching Patrol No. 2, the solutions Rich devised for his multiple problems also reflect the same pioneering spirit exhibited by the other men—and women—who volunteered for the assignment. Their "can do" attitude ultimately insured the operation's success. When no government reimbursement checks were received during the remainder of 1942, Rich helped negotiate personal loans among the staff members to keep them solvent. And when his pilots (who served without pay) found their $8.00 per diem fell short of El Paso's inflated housing costs, he leased temporary living space in the Alameda Auto Court. Once the patrol flights began Rich had to face another critical problem: fuel purchases. He discovered there was "no money with which to buy gasoline and no credit. It was, therefore, necessary that this expense be borne by the plane owners, but shortly thereafter we made our own contracts with Standard Oil Company of Texas who did extend us credit."[8]

---

7. Operations Directive no. 32, November 10, 1942, CAP Records.

8. Although the SLP volunteers received no salary and had to purchase their uniforms, they were compensated for the use of their personally owned aircraft. This was based on the aircraft's horsepower and covered operation, maintenance, depreciation, and insurance. Compensation ranged from $5.26 per hour for aircraft in the fifty to sixty horsepower range to a maximum of $37.26 per hour for the 400 to 445 horsepower range (Operations Directive no. 32, November 10, 1942, CAP Records).

Major Coffey must have established some kind of a wartime military record in launching the CAP Southern Frontier Liaison Patrol. In less than one month after issuing his final report, the civilian patrol planes were flying the United States–Mexican border. The Laredo base began operations on October 3, followed by the El Paso base on October 11. And once airborne, the dedicated civilians who volunteered their service in the patrol added an important chapter to the ongoing saga of men with wings that watched over the international border. They constituted a romantic quasi-flying circus, whose literary sobriquet, *flying minute men*, projected in historical perspective their contribution to national defense.

During the first three months of operation—October 3 to December 31, 1942—the patrols recorded 3,010 hours of flying time accumulated on 455 routine missions and 208 special missions. With 125 personnel maintaining and dispatching thirteen aircraft (eleven others were listed as out of commission), the flight crews reported sightings of twenty-one suspicious aircraft, eighteen suspicious signals or markings, and 173 unusual activities. This was achieved with six forced landings, one aircraft totally lost, and no injuries or fatalities (weekly telegraphic reports, CAP Records).

The airplanes that constituted the Southern Liaison Patrol fleet represented a mishmash of civilian airframe designs. There were Stearman and Waco (pronounced Wahco) open cockpit biplanes, a Fairchild 24, a Lambert-powered Monocoupe, Stinson monoplanes (the three-place "Station Wagon" and the four-place "Reliant"), a Bellanca "Cruisair," Carl J. Turner's staggerwing Beechcraft model D-17, George Copping's huge Rearwin biplane (known affectionately as the "Gremlin Castle"), plus several types of two-place tandem and side-by-side cub-type trainers.

The professional backgrounds of the flight personnel were as diverse as the planes they flew. Most were amateur pilots from various walks of life. In addition to a Secret Service agent and a commercial pilot, the flight roster included a flight instructor, a ship's captain, an aircraft company representative, a jukebox operator, a crop duster, a crane operator, a newspaper reporter, a real estate agent, a civil engineer, a manufacturing dentist, an attorney, a movie director named Henry King, and a movie actress, Mary Astor. King, who owned and flew a Jacobs-powered Waco biplane, worked in the A-4 Intelligence Section at the Brownsville base, while Astor, visiting a friend at the same location, worked temporarily in the communications center (a ground assignment).

Probably the most unique character in a unique cast of characters was Lt. George W. Copping. This rare individual had lost both legs

Flight line, Civil Air Patrol aircraft at Southern Liaison Patrol Base No. 2 headquarters, Biggs Field, Texas. Aircraft types from right to left are Beech-craft model D-17, Waco model "C," two Bellancas, and a Luscombe. Other aircraft not identifiable. Courtesy Richard A. Hagmann.

"Light planes were ideal for border patrol. Pilots hedgehopped along creek beds and ravines . . . [searching] for fugitives who might be hiding during the daylight hours." This SLP two-place Funk model "B" hovers over a rugged section of the Big Bend country searching for illegal trespassers. Courtesy Robert Neprud.

due to frostbite and was fitted with artificial limbs. When a friend told him about the ease of flying, he began taking lessons, purchased a Taylorcraft, and later became an agent for that company. Major Rich reported that during the patrol "Copping flew better than 1,000 hours . . . in all kinds of weather and under the most trying conditions." On one occasion he made a forced landing in an alfalfa field, "then walked eight miles to a phone in spite of his artificial limbs" (personal narratives, ibid.).

Lt. Dale E. Nolta, who operated an agricultural flying service at Willows, California, was typical of many of the civilian pilots who volunteered for the patrol. Professional training and experience equipped him well for the assignment. He, with his brother Floyd, a World War I pilot, formed the Willows Flying Service in 1928. Flying a converted Curtiss "Jenny," they pioneered aerial rice seeding in the upper Sacramento Valley. The business prospered, but with the outbreak of World War II, both brothers wanted to enlist in the Army Air Force. "They tossed a coin and Dale lost," Mrs. Nolta remembered. "He stayed home and we ran the business and Floyd joined the Air Force. The government thought seeding rice was just as important as dropping bombs" (interview, May 29, 1981).

When Nolta, a member of the local CAP wing, learned of the new patrol, he volunteered for the assignment. With most of the company's agricultural flying restricted to the spring and summer months, he accepted on a seasonal basis. "For two or three winters Dale went to El Paso, lived in a hotel, and flew with the CAP," Mrs. Nolta explained. "He had a strong sense of duty. He wanted to serve and this was his only opportunity." She could recall little of his patrol experiences, adding that "he was very secretive about his work there. Said they made some unauthorized flights into Mexico, but said that he never would talk about it. And he never did" (ibid.).

Because of the rugged terrain over which Nolta and his colleagues flew, the aerial smorgasbord that assembled at El Paso and Laredo was far better adapted for the surveillance assignment than the Air Corps units they replaced. The program outlined by Major Coffey provided for a low-level, close-in inspection of every foot of the border, searching for evidence of illegal crossing—footprints, hoofprints, tire tracks, or any indication of human movement across the border. This required a specialized skill in low-speed, low-altitude maneuvers. According to Civil Air Patrol historian Robert E. Neprud, these low-powered planes

> were ideal for border patrol. Pilots hedgehopped along creek beds and ravines, enabling observers to comb the brush on the banks

for fugitives who might be hiding during the daylight hours. When an automobile was spotted careening down obscure back roads toward the International Boundary, CAP crews sometimes flew low enough to read the license plates. (Neprud, 1948: 82)

Col. Frank Hubbell, chief of the Army Air Force Intelligence Section, A-4, at Biggs Field, who processed the patrol's intelligence data, recognized early on that, while his civilian colleagues encountered no enemy aircraft, the operation was nevertheless hazardous. "They flew twenty to fifty feet off the ground, always remaining very, very low," he explained. "They took quite a few chances. If you know that Big Bend country and those canyons, there's not much room for a slip" (interview, April 13, 1980).

Lt. Earling Felland, flying a low-level patrol directly over the Rio Grande, almost made the fatal slip. At less than three hundred feet his engine quit. Mexico was on one side, the United States on the other, but at that altitude the river was his only alternative. Preparing for a crash landing, Felland noticed a sand bar on the Mexican side of the river, and by tricky maneuvering sat his little Monocoupe down "where no plane had ever landed before." When Major Rich learned that Felland's only problem was water in the fuel tank, he began making plans to salvage the valuable aircraft. Faced with the additional problem of removing the aircraft from Mexican soil, Rich turned as a last resort to "Yankee ingenuity [which] did the trick." They removed everything possible from the aircraft to reduce weight and drained all the fuel except about three gallons. While Felland revved up the engine to gain power, the crew held the vibrating little craft stationary as long as they could before releasing it. The Monocoupe leaped forward, ran the full length of the sand bar, and at water's edge struggled into the air. "The landing and take-off . . . were things you read about but seldom [see] happen," Rich said of Felland's airmanship. ". . . The entire sand bar was less than 500 feet long!" (personal narratives, CAP Records).

In spite of unpredictable winds, thunderstorms, snow and ice, and occasional engine failures, the Southern Liaison Patrol posted an enviable operations record. The El Paso base commander recalled with justifiable pride that this was achieved "to the consternation of Base Operations at Biggs Field, [as] hundreds of take-offs were made when all Army ships were grounded" (ibid.). In addition, this involved maintaining a daily four-flight schedule of almost seven hundred miles over some of the Southwest's most treacherous terrain. The El Paso base dispatched two flights daily in each direction—east and west— to Del Rio, Texas, and Douglas, Arizona. One flight departed in the morning and one in the afternoon. (The Laredo base operated two round-trip flights to Del Rio, but only one to Brownsville.) Former air force Sgt. Leigh Tanger, who kept the situation map for the base commander at Biggs Field, explained that the departure times were purposely staggered, as "no two flights were made at the same time, in the same direction, on succeeding days. This was done so that no one could plan a border crossing to avoid a scheduled patrol flight" (interview, April 15, 1980).

Members of the Southern Liaison Patrol recognized the value of local intelligence and began soliciting information from those sources. The Juárez stockyards is a case in point. When a shipment of Mexican cattle arrived in Juárez, someone from the squadron headquarters would meet the owners and talk to them while their cattle were clearing quarantine. Sergeant Tanger explained: "We would try to determine if they had any information about German and Japanese activities in the interior of Mexico." In addition, some ranchers living in isolated areas along the Rio Grande also became regular sources of information. Patrol pilots would frequently land at some remote ranch where they "delivered newspapers to people who normally wouldn't get a paper for a week's time. They befriended the ranchers, who in turn would filter back information . . . about questionable activities along the Rio Grande." He noted parenthetically, "Primarily, we were an intelligence operation" (ibid.).

As Patrol No. 2 reached full operational strength, Johnson's Ranch again became the focal point of Big Bend aviation activities. With the exception of the aircraft and men who flew them, little had changed at the airfield. Elmo Johnson remained the premier border watch, the ranch house still provided a haven for transient airmen, and fifty-gallon drums of emergency fuel gave added insurance to the pilots and observers who flew the border patrol.

The location, approximately halfway between the El Paso base and the Del Rio turnaround point some 490 miles downriver, added to the facility's appeal. Ground relief became important concerns for the crews of the slow-flying civilian planes; remaining aloft six or seven hours each day took its toll in human endurance. And since the Civil Air Patrol pilots could exercise considerable personal initiative while on patrol, Colonel Hubbell explained that "dropping in on the Johnsons for company, rest, and possibly a warm meal was well within their prerogative" (interview, April 13, 1980).

Lts. D. E. Nolta and R. Compagnon and WO William E. M. Noel registered at the Big Bend airfield on December 12, 1942, and returned the following day accompanied by Lt. Charles M. Simon. All were from the Biggs Field base and their entries are the first to appear in the field register after the patrol launched daily operations. Over the next three months the patrol planes landed fairly regularly at Johnson's Ranch. During December 1942, five aircraft bearing nine crew members registered there, and the following month the field recorded thirteen landings and twenty-four personnel. Most planes arriving at Johnson's Ranch carried two airmen, the pilot and the observer. However, one unexplained Biggs Field pairing registered at the field on March 27, 1943: "Lieutenant W. J. Yost—Laura May Brunton and Flipper-Dog." Apparently Laura May Brunton, whomever she may be, succeeded in breaking Col. Harry H. Blee's men-only barrier. (All of my attempts to locate surviving Southern Liaison Patrol pilots proved futile.)

The old adage—"the more things change the more they stay the same"—held true at Johnson's Ranch during the early 1940s. From the days of Air Corps Lts. Thad Foster, Hugh A. Parker, and William L. Kennedy to the era of Civil Air Patrol Majs. Charles B. Rich and Jack R. Moore and Capt. Carl J. Turner, the bond that existed between the airmen and the Johnsons remained firm and devoted. Major Rich explained this unique relationship:

> At Johnson's Ranch . . . we received requests to transport everything from medicine to chicken feed when winter rains had washed out their only road. Mr. Johnson was one of the few people who had the service of a private air freight line, but in return our pilots were always given every possible aid and consideration. (Personal narratives, CAP Records)

As the war progressed, persistent reports of Axis agents operating from airfields in northern Mexico posed both a challenge and a threat to the border security forces. By maintaining sanctuaries south of the Rio Grande, it was relatively easy for fast, high-flying aircraft to penetrate the Southern Liaison Patrol's "aerial curtain," rendezvous at some remote borderlands airfield, and quickly disappear back into Mexico. This theme, played out during World War II along the Rio Grande, was destined to become hauntingly familiar during other wars and in other years: Korea, 1951, and the Yalu River; Vietnam, 1972, and the Cambodian Communist sanctuary; Lebanon, Syria, and Israel, 1984; ad infinitum.

The threats of enemy infiltration facing Major Coffey and the Southern Liaison Patrol in 1942 were well documented. During his initial survey various intelligence-gathering agencies armed him with convincing data. The enemy was abundantly near; northern Mexico appeared to shelter an Axis air force in exile.

A. L. Aherns, chief inspector of the Immigration and Naturalization Service at Laredo, told Major Coffey he had information that three of seven American-built bombers procured during the 1938 Cedillo Rebellion were still based at Maíz, a Mexican town located south of Laredo. Their airworthiness should be determined, Aherns suggested, as he suspected they were involved in some of the clandestine border crossings reported recently in that area. While in Laredo, Coffey also gathered data from navy intelligence on a former World War I German officer named Van Knoop (or Van Knob), who owned a ranch some one hundred miles southeast of Boquillas Gap. From this location Van Knoop and his son-in-law, Gordon Bennett, an American citizen, operated three airplanes from a paved runway, a somewhat elaborate getup for a ranching operation.

In these border security discussions, both army and navy intelligence sources acquainted Coffey with the name of Carlos Arizemendi (or Arisméndez). He was described as an ace pilot and soldier of fortune who flew for Gen. Francisco Franco during the Spanish Civil War. More recently he was suspected of being involved in the movement of mercury in the Big Bend country (from the Terlingua Quicksilver District) and other mines in Mexico for export to Japan. According to an Eighth Corps Area intelligence report, "Mercury valued at 30,000,000 pesos has been delivered to Axis Agents from points in northern Mexico, and mercury worth about 7,000,000 pesos has been transported in the Axis-owned plane piloted by ARIZEMENDI" (CAP Records).

This evidence of enemy bases so near the United States border prompted Major Coffey to make a highly suspect recommendation that may or may not have been carried out. In his September 4, 1942, preliminary report, he stated: "Itinerant unidentified CAP personnel, without CAP markings on their ships, should make flights into Mexico, visiting airports where it is suspected that unscrupulous personnel base their aircraft" (ibid.).

While it is doubtful if this recommendation was ever enacted, the Southern Defense Command continued to monitor these adjacent enemy operations. A secret G-2 Periodic Report issued by that agency identified "a landing field, allegedly Japanese owned, located 90 miles west of Villa Acuna." A later report dated August 2, 1942, stated that the Southern Land Frontier had reliable information that "there is a large landing field near Patos, Durango, owned by a German whose name was not reported. This field is said to be equipped with landing light flares" (RG 338). Later that year, according to a newspaper account, "government agents investigating reports of a secret Axis airfield in Chihuahua state, captured six Japanese, a German and an Italian" (*El Paso Herald-Post*, October 22, 1942).

Undoubtedly these remote Mexican airfields were serving as launching sites for unauthorized and unfriendly flights into the United States. Again the Southern Defense Command G-2 Intelligence Section supplied the evidence:

> Unidentified black monoplane observed at 10:50 AM crossing into U.S. from MEXICO in the vicinity of Columbus, N.M., and flying E. along border. Believed monoplane is property of GERHARDT HEIMPLE, NAZI suspect of CHIHUAHUA CITY. (March 3, 1942, no. 48)

> Fast, bi-motored unidentified plane flew northward over Lordsburg (N. Mex.) 5 June. Has been observed over that locality every week for a month. (June 6, 1942, no. 126)

> Civilian resident of Corona, New Mexico, reports that at 062006 Z, four unmarked airplanes had circled that area for 20 minutes. Source: Southern Land Frontier. (August 7, 1942, no. 188) (RG 338)

Concurrent with these reports, the air force intelligence section at Fort Bliss received a report of a remote landing field in New Mexico, supposedly being used by unauthorized aircraft coming from Mexico. According to former air force Sgt. Leigh Tanger, an intelligence team visited the site and "found several barrels of high octane gasoline and evidence that airplanes had recently landed there" (interview, November 29, 1980).

The combination of recurring sightings and reports of Axis airfields in northern Mexico led ultimately to the Southern Liaison Patrol's involvement in the matter. Major Rich's account of the investigation suggests that some of his civilian pilots exercised undue personal initiative in their pursuit of evidence. "Being closely allied with G-2," he wrote later, "we were called upon to make many observations on patrols, one of which concerned an airport near the border where planes of German make were supposed to land." They found the field and no aircraft was sighted, but with its location well south of the Mexican border, "we were unable to go over it because of State Department regulations. Col. [Vance T.] Batchelor [chief of staff of the Southern Land Frontier] and I both suspected that one of our pilots slightly violated the rule as he gave a very complete report on the field" (personal narratives, CAP Records).

The patrol's interest in the Mexican airfields stemmed in part from persistent rumors that airplanes were smuggling large amounts of United States currency—either confiscated or counterfeit—into this country to finance subversive activities. Major Coffey first learned of this operation from the Office of Naval Intelligence during his initial survey mission. The agency informed Coffey that

> last week [mid August 1942] . . . two millions in American money was landed in Mexico for the purpose of Sabatoge [sic]. . . . The Naval Intelligence . . . inferred that Carlos Armismendez [sic] may be involved [in transporting this money into the United States] and it was expected that he would not fly a ship in this type of work unless it had the proper armament with which to protect himself and cargo. (CAP Records)

From the description supplied by the Southern Defense Command, G-2 Section, a Civil Aeronautics Administration inspector identified the invading aircraft as a Junkers JU-88, a fast twin-engined medium bomber developed for the German Luftwaffe. According to the inspector, this specific JU-88 model cruised at approximately 238 miles per hour and carried about nine hours of fuel supply. This report, dated August 4, 1942, classifying both the source and the information as "reliable," concludes that "it must be presumed that this is an enemy aircraft, probably operating from Mexico into the United States from hidden airports" (ibid.).

Since the path of these reported over-flights fell within the scope of the Patrol No. 2 area, members of the civilian air force decided to enter the chase for the mystery craft. (On several occasions the air force reportedly dispatched fighter aircraft in unsuccessful attempts to intercept and identify the elusive airplane.) They faced one major obstacle, which Bertrand Rhine, California Civil Air Patrol wing commander, explained:

> They had intelligence that a German airplane flying at a very high altitude, and very fast in those days, was transporting money—or counterfeit money—and landing some place north of the border. And they wanted to try to intercept that airplane but didn't have anything that was fast enough to do it. (Interview, May 13, 1980)

Rhine knew of a privately owned aircraft that he thought possessed the performance capability equal to the assignment. Paul Mantz, a West Coast charter service operator and motion picture stunt pilot, owned a specially equipped Lockheed "Orion" monoplane, which he had entered in the 1938 Bendix Trophy Race. (Mantz placed third at an average speed of 206:579 miles per hour.) Rhine stated he recruited the aircraft and

ferried it to Biggs Field in El Paso, Texas. It was used on our border
patrol base . . . and was originally intended to intercept a German
. . . airplane that had been seen crossing the border repeatedly.

The aircraft was equipped with a large . . . engine . . . and special gas tanks. I might add that the [fuel] plumbing on the airplane
was a plumber's nightmare because of all the tanks. . . . [But] this
plane was a real delight to fly as of the period of World War II.
(Correspondence, December 8, 1980)[9]

While there is no evidence that the mysterious high flying aircraft was ever apprehended, United States currency in Axis control
continued to circulate in the Western hemisphere, much of it being
smuggled into the United States. The United States Treasury Department took action in the matter. On August 14, 1942, Secretary of the
Treasury Henry Morgenthau announced that

the Government of Mexico and the Government of the United
States have taken steps to . . . [prevent] the disposition within the
Western Hemisphere of currency looted by the Axis.

The Government of Mexico has prohibited the importation
into that country of all United States currency other than bills of
two-dollar denomination and United States coins. (Department
of the Treasury, general ruling no. 14)

Designating the comparatively rare two-dollar bill as the only currency that could be legally crossed at the international border apparently forced the Axis agents to turn to high-speed aircraft to transport
the illicit funds. And while Paul Mantz's supercharged Lockheed
"Orion" failed to intercept the elusive aircraft, its presence did hinder
the enemy's effort to penetrate United States air space.

By mid 1943, through the combined efforts of all border surveillance agencies, earlier fears of enemy aggression through Mexico had
gradually subsided. Nowhere was this more evident than at Johnson's
Ranch; during the summer and autumn air traffic virtually ceased.
From March through August—excepting June—only one flight each
month registered at the field. No clue to this change appears in the
field register, nor does Neprud note in his work any administrative or
operational changes in patrol policy. Also, no flights appeared at the

---

9. When developed in 1930, the Lockheed "Orion" was the first commercial airliner in the world to exceed two hundred miles per hour and "was the
first on which flaps were used to reduce landing speed and increase the angle
of descent" (Davis, 1972: 177).

ranch in either September or October, and only three flights make up the November traffic. On November 5, 1943, a first lieutenant (signature illegible) from the Biggs Field base registered solo at Johnson's Ranch. A trio made up of 1st Lt. Lewis H. Hanson, 2d Lt. Ted G. Misenhimer (both from the Biggs Field base), and E. C. Metz, a Civil Aeronautics Administration representative from the Fort Worth regional office visited the airfield on November 11, 1943.

One entry remained to complete the month's traffic. On November 25, 1943, 1st Lt. Harold C. Filbert entered his signature in the Johnson's Ranch field register. Unknowingly, he inscribed for himself a small niche in the history of aviation along the nation's Southwest Borderlands. His is the final signature in the field register. The lower half of page twenty-one was destined to remain blank. When Lieutenant Filbert's plane lifted off the runway at Johnson's Ranch on that November day, it signaled the final closing of a facility that for fourteen years—from the Escobar Rebellion to World War II—brought a measure of safety and security to that remote section of Texas.

As the little patrol plane cleared the field and began winging its way across the Big Bend wilderness, it left behind almost a decade and a half of memories and traditions. The World War I De Havilland DH-4s that first shattered the pristine silence of the Big Bend country were followed by a seemingly endless wave upon wave of military aircraft, many bearing some improvement over the ones that had come before and all pointing to that day in the future when they would lead the nation's forces into battle. The future was now. Many of the young lieutenants who once flew the little biplanes with the bright orange wings to Johnson's Ranch were unknowingly "learning how to fight one hellova war" (Kennedy, interview, January 19, 1978).

The young men who first signed the field register at Johnson's Ranch were symbolic of an age when men with vision and imagination were building an air force and training people to insure its effectiveness. These were the members of the "Order of the White Scarf," those who emerged in leadership roles in the nation's air armada. Their Big Bend experiences had served them well. By November 1943, history had already recorded their deeds in battle; Fortress Europe and the Japanese homeland still lay ahead.

Where were they on that day in late November 1943? Lt. Joseph H. Hicks, who once strafed targets behind Elmo Johnson's trading post in a Thomas-Morse O-19, was a major serving as chief of supply in Hollandia, New Guinea; Lt. Don Mayhue, who flew Elmo Johnson on an emergency mission to a hospital, was a colonel on another emergency mission with the Ninth Air Force in England. His target—Germany. Lts. Elmer P. Rose and Hugh A. Parker, both recently promoted to

Transfer of command, Southern Liaison Patrol, November 1, 1943, at Biggs Field, Texas. Maj. Charles B. Rich (far left) relinquishes command to Maj. Jack Moore, while Maj. Frank A. Hubbell, United States Cavalry intelligence officer, and Capt. Charles S. Simon, CAP executive officer, look on. Lt. George W. Copping, the CAP pilot who was missing both his legs, is visible in the rear. Courtesy Richard A. Hagmann.

Lt. George W. Copping, with his back to the camera, files a patrol intelligence report at the Southern Liaison Patrol Base No. 2 operations office at Biggs Field, Texas, while Capt. Carl J. Turner (wearing cap), operations officer, looks on. Operations board is in background. Courtesy U.S. Army Signal Corps and Col. Thomas H. Daniel, Jr., CAP.

colonels, held administrative positions in the Pentagon in Washington, D.C.; Col. Thad V. Foster, following an assignment with the Alaskan Defense Command, was serving as base commander at the Jackson Army Air Base in Jackson, Mississippi. Another recently promoted colonel, LeRoy Hudson, was base commander at Tinker Field in Oklahoma City. Maj. Nathan F. Twining, now a brigadier general, commanded the Thirteenth Air Force in New Caledonia. The wheel of fortune played vastly different games for Lts. John W. Egan and William L. Kennedy. Both colonels, Egan commanded a navigation training base at Selmon Field, Monroe, Lousiana, while Kennedy, also a base commander and armament specialist eager for combat, signed on as an observer on an ill-fated bombing mission over Germany. He remained for the duration at Stalag Luft Three. Fate denied him the opportunity to fight that "one hellova war" he had spent more than a decade preparing for.

The closing of Johnson's Ranch also marks the beginning of the end of the Southern Liaison Patrol. Operations continued well into 1944, but "the continued lessening of enemy capabilities" prompted a careful reevaluation of the nation's entire southern defense perimeter, which included the Southern Land Frontier. A telephone call from the Operations Division, War Department, in March 1944 "requested our [Southern Defense Command] reaction toward the discontinuance of the Civil Air Patrol along the Mexican border." Maj. Gen. Henry C. Pratt, recently appointed SDC commanding general,

> agreed that the cost of operating these reconnaissance patrols was disproportionate to the military benefits gained therefrom. [He] . . . therefore suggested that if G-2, War Department concurred in his estimate that there was little danger of aggression from or through Mexico, the Civil Air Patrol be discontinued. As a consequence, Headquarters Army Air Forces issued a directive to discontinue the Civil Air Patrol, and operations were terminated on 10 April 1944. The saving effected thereby, as shown by a previous study by the G-3 Section, was an estimated $643,000 annually.[10]

This marks the end of the final chapter of one of civil aviation's major contributions to national defense during World War II. The final summary of operations, covering the entire period from October 3, 1942, through April 10, 1944, reflects the magnitude of the operation.

---

10. G-3 *Supplement, Historical Record, Southern Defense Command, for Period 25 January–31 December 1944*, p. 46, RG 338.

```
Hours flown. . . . . . . . . . . . . . . . . . . . . . . . . . . . . . . . . . . . . 30,033
Routine patrols flown  . . . . . . . . . . . . . . . . . . . . . . . . . . . . 4,720
Special missions flown. . . . . . . . . . . . . . . . . . . . . . . . . . . . . 1,379
Suspicious aircraft observed and reported . . . . . . . . . . . . . . . . 176
Suspicious signals or markings
        observed and reported  . . . . . . . . . . . . . . . . . . . . . . . . . . 436
Unusual or out-of-the-ordinary activities
        observed and reported  . . . . . . . . . . . . . . . . . . . . . . . . . 6,874
Forced landings. . . . . . . . . . . . . . . . . . . . . . . . . . . . . . . . . . . 54
Airplanes lost . . . . . . . . . . . . . . . . . . . . . . . . . . . . . . . . . . . . 13
Fatalities . . . . . . . . . . . . . . . . . . . . . . . . . . . . . . . . . . . . . . . . 2
Personnel seriously injured  . . . . . . . . . . . . . . . . . . . . . . . . None
```
(April 15, 1944, CAP Records)

With varying lengths of enlistment, personnel totals fluctuated from week to week. The staff averaged around 130 and operated some thirty-five aircraft, but seldom more than half that number were operative. During the week of February 5, 1943, personnel dropped to 106, operating fourteen of twenty-seven assigned aircraft on sixty routine missions and flying 352 hours. One high point of the operation occurred during the week of November 19, 1943, when both patrols logged 533:30 hours on seventy-five routine patrols and thirty-four special missions, during which they reported three suspicious aircraft and 135 unusual activities. This was achieved with 150 personnel with fifty assigned aircraft, half of which were operative (weekly telegraphic reports, CAP Records).

Two categories, "airplanes lost" and "fatalities," reflect the moments of high drama and tragedy experienced by members of the Southern Liaison Patrol. On the night of March 12, 1943, a fire in the hangar and shop of Patrol No. 2 at Biggs Field destroyed "nine much needed ships. . . . This nearly wrecked our whole operation and took years off the Commander's life" (Rich, personal narratives, CAP Records). Three other aircraft were reported lost to operational accidents. There remains, however, the matter of the one lost aircraft and the two fatalities. This will probably remain forever the one unexplained tragedy of the Southern Liaison Patrol.

The Weekly Operations Report of January 22, 1944, states briefly and succinctly: "One aircraft was totally destroyed and both pilot and observer were killed as a result of a forced landing" (CAP Records). No other description or explanation of this tragedy appears in the operational records of the Southern Liaison Patrol. The CAP fatality report issued on August 8, 1947, more than three years later, offers no clues to the cause of the accident: Stinson aircraft NC 27734 piloted by H. M. Hewitt crashed seven miles from Guerrero, Mexico, while on routine border patrol (ibid.). Only one brief newspaper account and the recollections of contemporaries shed some light on the possible cause of the crash. On January 19, 1944, *three days after the accident*, a Fort McIntosh public relations officer announced that Lts. Harry Hewitt and Bayard Henderson were killed while "on a routine patrol twenty miles from Eagle Pass [Texas]. . . . The plane crashed and burned" (*Laredo Times*, January 19, 1944).

According to Col. Frank Hubbell, former Biggs Field intelligence officer, "the matter was really hushed up." He ordered an investigation of the crash but was told that "San Antonio [Southern Defense Command headquarters] would handle it. Really teed me off, but that's the way it goes sometimes in the military. The gossip was it was a gunnery school accident." Retired Sgt. Leigh Tanger recalled the same story, adding that "the whole thing was further complicated, as I recall they went down on the Mexican side of the river" (interviews with Frank Hubbell and Leigh Tanger, January 11, 1984).

Lieutenant Hewitt's widow was equally unsuccessful in gathering the facts about her husband's death. She explained the death certificate issued by a Del Rio mortuary remains unsigned, the cause of death omitted. Bits of information provided by the lieutenant's colleagues also supported the gunnery school theory. "They suspected it was from live ammunition fired from one of our planes out there," Mrs. Juanita [Hewitt] Dichter recalled some four decades later. "But it was never proven" (interview, February 19, 1984).[11]

In the absence of an official account of the accident, the newspaper report, the personal recollections, and the location all point to Frank Hubbell's explanation—"a gunnery school accident." The Eagle Pass Army Air Field, an Army Air Force advanced training facility, maintained a gunnery range south of the field and near the reported crash site. In a training maneuver, student pilots would fire live ammunition at a target towed behind another aircraft. It is possible that projectiles fired at high altitude at the moving targets could indeed have accidentally struck the low-flying patrol plane, causing the fatal accident. The military, however, keeps its secrets well; the facts will probably never be known.

---

11. Gaining survivor's benefits proved to be as difficult as determining the cause of the accident (Harry Mark Hewitt Dichter was born twenty days following his father's death). Both houses of the Oregon legislature passed a joint resolution supporting Mrs. Hewitt's insurance claim, and this led to a similar bill being introduced in Congress. Some two and one-half years later her claim was approved before the bill ever came to a vote.

The fifty-four forced landings reported by the Southern Liaison Patrol produced various results. Some, like the Fairchild 24 (above), came to rest right side up. Others such as the Luscombe (below) inspired one of the patrol's more ludicrous sobriquets: the "upside-down air force." Courtesy Richard A. Hagmann.

This single fatal crash, occurring during 30,033 hours of flying over some of the Southwest's most treacherous terrain, should not detract from the patrol's achievement. The foregoing statistics speak for themselves. When considering this was the undertaking of voluntary civilian personnel who provided all the aircraft and much of the equipment, these statistics assume even greater importance.

Although the patrol by its very nature maintained a low surveillance profile, there was both immediate and tangible evidence of its success. Major Coffey stated in a report to Lt. Col. Earle L. Johnson, CAP national commander, that with the advent of the SLP, suspicious aircraft sightings in the Western Sector had been reduced drastically, and radio reports of land crossings "brought about the apprehension of individuals almost immediately." In addition to providing an aerial courier service for the military—mail, messages, and reports—Coffey noted that the patrol's aerial surveillance made possible substantial troop reductions all along the United States–Mexican border (October 4, 1943, CAP Records). In addition to acknowledging the economic benefits accruing to the military, Gen. W. B. Augur, commander of the Southern Land Frontier, paid high compliment to the personnel that constituted the civilian air force. "They . . . would be an asset to any organization . . . [and will] do everything possible in the fulfillment of their mission" (July 18, 1943, ibid.).

Maj. Charles B. Rich likewise saw the assignment as a personal commitment to a high cause. He wrote later:

> The Operation brought out heroic action which will never be recognized by the Army or known to the people. Those pilots who flew, day after day, over one of the toughest terrains in the United States, in planes which were in poor condition, under all kinds of weather [conditions], at an altitude between 250 and 500 feet, performed a service as heroic as any Air Force man in Service. All of them should receive the Merit Award. (Personal narratives, CAP Records)

The official account of the termination of the Southern Liaison Patrol, brief in its phraseology, terse in its manner, nonetheless contained important overtones undoubtedly overlooked by the military command. Implicit in this announcement is the passing of two significant milestones, one historical, one cultural. Both had far-reaching significance within the contemporary American experience.

First, the position of the Republic of Mexico as an ally in the ongoing conflict as well as that country's "extremely cordial" relations throughout the duration of the patrol operation lend further evidence that the Escobar Syndrome had finally run its course. And while those latent fears might still exist in some quarters, they never again reached overt expression. Such bogus currency was no longer negotiable.

And, like the passing of the Escobar Syndrome, the demise of the "Order of the White Scarf" was destined to follow. Although the civilian crews that flew the little patrol planes along the Rio Grande scored no aerial victories, they nevertheless helped "shoot down" the romantic myth of the "intrepid flyer," an image born in the skies over Europe a quarter-century before. The civilian pilots that "landed on sandbars in the Rio Grande, made the repairs themselves . . . and flew 'em out" were a far cry from the likes of Lt. Thad Foster, who relished the company of his female entourage while basking in the spotlight of public adoration, or like Lt. Charles W. Deerwester, who, with Elmo Johnson as his passenger, used Air Corps planes to chase suspected murderers into Mexico mostly because they both "just liked to play airplane."

The pilots of the Southern Liaison Patrol were a far more pedestrian group who flew their own airplanes, did their own maintenance, and serviced their own equipment as part of their volunteer commitment to their country. To these pilots, aviation was a business; romanticism was something they had read about. Within their world slowly but surely the airplane had ceased to be an oddity. Their ever-growing numbers, many flying the little planes with less horsepower than the automobiles they drove, were altering the course of American aviation. The specter of the "Order of the White Scarf" was gradually being eroded away.

Mid-century realism reserved no space for romantic myths. Gone was the glory that accompanied Eddie Rickenbacker as he patrolled the Western Front, Jimmy Doolittle as he cornered pylons in a stubb-wing racer, or Thad Foster as he followed the Southern Pacific railroad tracks in an old DH-4 biplane to establish a Big Bend airfield. Johnson's Ranch, while serving military aviation for a decade and a half, was, in essence, a landmark.

"... I would like to say again how much we appreciated and enjoyed the Johnsons"

CHAPTER 17

The closing of the former Southwestern Airways Station appears symbolic of changes occurring in the personal lives of Elmo and Ada Johnson. A way of life they had enjoyed together beside the Rio Grande was fast disappearing. They would soon follow in the wake of the departure of the Air Corps and the Southern Liaison Patrol.

For almost two decades the Johnsons had lived the lives of modern pioneers, prospering within that strange twilight zone between two divergent worlds. Just west of the ranch house and across the river lay Mexico's northern frontier. People from this primitive land came to regard the Johnsons as friends and guardians of an economic oasis that enriched their way of life. North and west of the trading post stretched the runways of the landing field, where modern military aircraft brought the outside world to their riverside doorstep. And from this strangely unique posture, positioned midway between these two worlds—the primitive and the contemporary—Elmo Johnson remained the dominant figure, both by personal stature and military decree: "No rank will be observed at Johnson's Ranch, except for Elmo, who is always in charge."

The decade of the 1940s brought many changes to the Big Bend. Probably nowhere were these more apparent than at Johnson's Ranch. Long before the Air Corps abandoned the landing field, the Seventy-fourth Congress had authorized the establishment of a national park in the Big Bend area. This had been an enduring dream of Ada Johnson, who had widely promoted the idea. Both she and Elmo had envisioned a park that would preserve the region's primitive lifestyles, where their ranch and trading post would be maintained as part of the park's interpretive concept. But this was not to be. Ada Johnson would never realize the fruits of her labor.

In 1941, the Forty-seventh Legislature appropriated $1.5 million to purchase land for a national park. Acquisitions began on September 1, 1941, and on April 1, 1942, Elmo Johnson received Warrant No. 74 in the amount of $12,599.83 as payment for 996.3 acres of land. The land was valued at $492.43 and the improvements at $12,007.40.[1] The trauma of selling the Big Bend ranch was aggravated further by a violent personal and legal conflict between Johnson and a tenant who lived on the ranch at the time of the sale.

According to former Border Patrol inspector George W. Harrison, one Louis Lewenthal, along with a son and a daughter, emigrated from "somewhere up north and settled on the Johnson ranch" (interview, November 26, 1980).

The rental agreement between the two men, obviously verbal, is not clarified in the legal documents relating to the matter. Lewenthal claimed, however, that at various times between March 7, 1939, and September 13, 1940, he paid Johnson $610 "in rental prepayments of the rental value of the premise occupied by defendant." The payments, Lewenthal states, consisted of $605 in cash, plus a dog valued at $5.[2] The relationship between the two men obviously deteriorated and on January 15, 1941, "the defendant [Lewenthal] unlawfully entered upon said premise and ejected plaintiff [Johnson] therefrom" the disputed ten-acre parcel of land unidentified in the legal documents. It is understandable that Lewenthal, an outsider unfamiliar with the traditions of frontier society, may have thought that forcibly ejecting Elmo Johnson from the property was in his best interest. But if Lewenthal believed this, he quickly learned otherwise. Johnson still believed that the best and most expeditious law of the Big Bend was the one most people carried in a holster or slung over their back. In the ensuing encounter, Lewenthal claimed that both Ada and Elmo Johnson shot at him repeatedly, an accusation Johnson never denied. In an oblique justification of Lewenthal's survival, Johnson, the ultimate marksman, explained, "I never shot to kill, though. I never had to, and I wouldn't have wanted to" (Cox, 1970: 62). His method still proved viable; the disputed issues were resolved on April 7, 1942. Johnson agreed to pay Lewenthal three hundred dollars and Lewenthal relinquished "all of his right, title, and interest in, and possession to the lands hereinabove mentioned."

---

1. Texas Parks and Wildlife Department, Records Center.

2. *Case No. 2014*, Elmo Johnson, plaintiff versus Louis Lewenthal, defendant; Louis Lewenthal, cross-plaintiff versus Elmo Johnson, Brewster County, and Works Project [sic] Administration, cross-defendants (vol. 99, p. 179, District Court, Brewster County, Alpine, Texas).

On September 5, 1943, some three months before the last airplane departed Johnson's Ranch, Texas governor Coke R. Stevenson presented the deed to 788,682 acres of land to M. R. Tillotson, director of Region III, National Park Service. This primitive wilderness area was to become the Big Bend National Park. On June 12, 1944, when the park was established officially, neither of the Johnsons was present for the dedication.

Johnson had once envisioned converting the trading post into a tourist attraction and, according to Gen. William L. Kennedy, the airfield, the radio station, and the officers quarters all seemed to fit into this plan. The approval by the Works Progress Administration of a $10,086.72 expenditure to renovate and expand the existing auxiliary structures together with the ultimate development of a national park on the property gave added plausibility to the scheme (RG 237). The project had been completed in March 1936, and visiting airmen and other state and federal employees used the facility intermittently during the late 1930s. As air traffic to the Big Bend airfield declined with the coming of World War II, the rooms of this motel-like structure, standing vacant and empty (and sometimes filled with surplus farm produce), seemed to foretell the approaching eventide of Elmo and Ada Johnsons' Big Bend tenure. Sometime later a rare windstorm destroyed the water tower, erected as part of the WPA project to serve the radio station and the officers quarters. Both the tank and the steel tower were left in shambles. "I always felt from what I heard," wrote David L. Smith, "that was considered the last straw of the successful operation in the Big Bend" (correspondence, November 18, 1980). With Ada Johnson's deteriorating health they abandoned all hope of remaining with the National Park Service. Sometime in late 1943 they moved to Sonora, Texas, to live on property Ada Johnson had inherited from her father.

The last years were not pleasant for Ada Johnson. In addition to leaving the region and the friends for which she held such deep devotion, she was waging a losing battle with cancer. There were several operations, the first performed in San Angelo in 1941. When Lt. Don Mayhue, the Johnsons' old friend, then stationed at Brooks Field, learned of her illness he flew to San Angelo to visit with her in the hospital. He remembers she asked about their old Air Corps friends and talked about the good times they had enjoyed together at the ranch. At that time, Mayhue recalled, she had high hopes for a complete recovery from the disease. Her optimism, however, was ill-founded; there were other operations, but she grew steadily weaker, and on July 3, 1945, Ada Johnson died.

Elmo Johnson remained in Sonora, where he operated a woodworking shop and a gun shop, built boats, and maintained some rental property, the latter becoming his primary source of income. Around 1955, Johnson gave up his other enterprises and opened the Corral Trading Post in Sonora, a sporting goods and western gift shop. The Corral prospered for a while, but Johnson soon discovered that his two partners brought to the business the seeds of economic failure. "They were two of the biggest sporting goods purchasers in the area," Smith wrote, "who thereafter made all their purchases, and arranged most of their friends purchases, at cost." Expenses mounted faster than receipts and Johnson went deeper and deeper into debt. Subsequently he sold the home on the courthouse square in Sonora in order to keep the Corral Trading Post solvent. Apparently this was too little too late. Johnson's debts finally reached the point that he could no longer obtain merchandise on credit, and he was forced to close the business.

Following the sale of his home in Sonora and the failure of the Corral Trading Post, Johnson moved to a little two-room house on his brother's ranch on the outskirts of Sonora. Here he spent his remaining years alone, mostly taking care of his rental property, gardening, and feeding and protecting wild game that roamed near his house. Although Johnson's thoughts undoubtedly were drawn back to his life along the border, he seldom mentioned the past. During his last years "he steadfastly refused to talk about . . . [the Big Bend]," Smith remembered. "He would always say, 'I'm not interested in talking about the past. I'm interested in the future'" (ibid.).

But the past remained very much with him. Following the end of World War II many of the Air Force pilots who had once landed at the Big Bend airfield came to Sonora to see him and to talk about the good times they had enjoyed together along the Rio Grande. Until the time of his death he continued to receive "cards and letters from people from all over the world. People whose names I don't remember, and wouldn't know 'em if I saw 'em" (interview, June 3, 1966). One he did remember with great pride was Gen. Nathan F. Twining, by then chairman of the Joint Chiefs of Staff, who wrote Johnson:

I have just received from Mr. W. D. Smithers of Alpine some pictures taken of the old book [field register] from Johnson's Field. This thoughtful gesture on his part certainly brings back some of the most pleasant memories I have of Texas and the old Air Corps people of those days, to say nothing of the hospitality of Mrs. Johnson and you. I shall enjoy having these, and I think it is only fitting that the book go to Sul Ross State College for preservation of this bit of tradition.

I am sorry to hear of Mrs. Johnson's passing away. However, I do hope that this finds you in the best of health and that things are going well for you.

I would like to say again how much we all appreciated and enjoyed the Johnsons. (Letter dated March 1, 1958, copy in my possession)

During the last two decades of his life, Johnson returned only twice to the Big Bend. His niece, Mrs. Emerald Smith, believes he made the first trip shortly following Ada's death. Mrs. Smith and a Mr. and Mrs. Ed Mayfield of Sonora accompanied him on this occasion. Originally, they planned to visit the ranch site; however, they altered their plans after reaching Castolon. Some old friends at the Castolon Trading Post told him that recently park employees had bulldozed most of the ranch buildings.[3] The party returned to the Chisos basin tourist center where they spent the night. Unexpectedly, the visit developed into a homecoming of sorts for Johnson. Mrs. Smith remembered "quite a to-do was made for [him] . . . at a gathering of park visitors in the evening" (correspondence, November 18, 1980).

Johnson returned to the Big Bend once more and this time he visited the old ranch site. In early December 1974, he visited his daughter, Mrs. Thelma Ashley, in San Antonio, where he purchased a new Ford pickup truck. After acquiring the pickup, he suggested that he, Thelma, and her husband, E. G. ("Buster") Ashley, drive out to the Big Bend. He said that he felt great and wanted to do some traveling. Mrs. Ashley attempted, with no success, to discourage the trip. "Thelma didn't want to take him down there. Didn't want him to see what was left there," Mr. Ashley recalled. "He kept on urging her and she finally agreed to go with him" (interview, December 5, 1980).

They arrived at the Big Bend National Park visitor's center late and spent the night there. Ashley recalled that during the night Johnson had difficulty breathing at that high altitude (almost eight thousand feet) and they thought he was having a heart attack. The following morning, however, he said he felt able to continue the trip.

When they arrived at Castolon, Johnson stopped at the old trading post and appeared disappointed when he recognized no one there and no one knew him. That was a premonition of what lay ahead. On leaving Castolon, Ashley said that he felt that Johnson began having some misgivings about his decision to return to his former home site.

---

3. This did not include the ranch house. According to Dr. Ross Maxwell, former superintendent of the Big Bend National Park, after Johnson moved local residents removed the roof, windows, and all wooden parts of the structure. The unprotected adobe walls gradually melted in the late summer rains.

A 1962 aerial view of the ruins of Johnson's Ranch looking south toward Mexico. Ranch house and trading post, center; radio station and officers quarters, right center. Courtesy Smithers Collection.

His suspicions were soon confirmed. As they drove past what was once the airfield and approached the spot where the home and trading post once stood, all that remained were three pillars of melting adobe brick that once supported the patio ceiling. Johnson parked the pickup and got out. He then walked a short distance, paused, and spoke. Not to anyone in particular, just spoke. "All my rock work [in the patio] is gone. Nothing left but a pile of dirt." As his daughter had anticipated, "he didn't expect to find the place like that" (ibid.).

After spending a few moments in silence, Johnson walked slowly back to the pickup, leaned against the hood, bowed his head, and began to cry softly. After a moment his daughter walked over to where he stood, put her arm around him, and gently helped him back into the cab. Silently they drove away. That was Elmo Johnson's last visit to the Big Bend. About three weeks later, on Christmas Eve, 1974 he died of a heart attack. He was eighty-five years old (ibid.).

"... my love affair with airplanes was over"

EPILOGUE

The Johnson's Ranch experience evolved as a military aviation time piece incident only to the Big Bend country of Texas during the 1930s. The convergence of optimum social, political, and technological developments led to the establishment of this remote ranchland facility. Created in the wake of Mexico's last major political upheaval, maintained to insure regional security, and ultimately abandoned when long-range aircraft and the exigencies of world conflict rendered it useless, the airfield developed as one of the United States Army Air Corps' truly off-beat installations. There was nothing else quite like Johnson's Ranch.[1]

The establishment and operation of this airfield also provide an insight into the status of military aviation and international relationships during the interwar period, the social and cultural phenomenon created by the emergence of a new cast of American hero-celebrities, and the unique posture they occupied in the minds and imaginations of millions who gazed skyward to the distinctive sounds of men in flight.

These were America's heroes of the air, members of the "Order of the White Scarf," who helped build America's aerial armada, trained its fighting men, and led them to victory in World War II. As in the previous world conflagration, it was the fighter pilots, those achieving ace

---

1. When examined in its total context, the Johnson's Ranch military experience was indeed unique. During the early 1930s, there were, however, other recreational areas where Air Corps pilots were authorized to visit in military aircraft. One such site was Grangerville (named for a Sergeant Granger), a hunting and fishing camp established by members of the Third Attack Group on Matagorda Peninsula some 110 miles west of Fort Crockett. The landowner provided a house and the Air Corps assigned several enlisted men to the camp to maintain and operate the facility.

status, that became household words—Bong, Boyington, Foss, and Gabreski. Tank commanders and destroyer skippers probably wreaked far greater havoc but seldom achieved comparable recognition. Call it adventure, romance, fascination with flying, or even patriotism, they helped develop aviation to potentials inconceivable when Thad Foster first landed his DH-4 at Johnson's Ranch. Their careers, accented further by the specter of the atomic bomb, jet aircraft, missile technology, Korea, and Vietnam, became milestones in the continuing history of combat aviation.

The jet age air force bears scant resemblance to the Army Air Corps that established Johnson's Ranch. Only the men (and women) and the mission provide the threads of continuity. The machines they fly, guided by radar, computers, and banks of electronic sophistication, function in a world far removed from the era of the DH-4, the O-19, and the B-10. Other than the material and equipment, the essential difference lies in the role of the individual and his or her personal initiative in the use of that equipment.

The escalating cost of jet aircraft, the fuel they consume, and the exacting demands of landing facilities have long since terminated the personal use of military aircraft. Strange field landings, navigation training, and the "just get out and fly" policy no longer fulfill military needs. Although in 1932 these were legitimate reasons for Lt. William L. Kennedy to fly out to a South Texas ranch in a military aircraft to shoot quail, such is no longer the case. Landing today's jet fighters on anything except a prepared surface (a paved runway) invites disaster.

With the jet age came the end of the weekend jaunts to New Orleans or other famous watering holes, once justified as cross-country navigation training. Today that skill means something entirely different. Since contemporary training flights are conducted at 540 knots (over six hundred miles per hour), "everytime you pull up the gear you are navigating," explained Gen. William R. Nelson, commander of the Twelfth Air Force. "There is no way you can stay around the home base [at that speed] . . . so you learn [aerial navigation] almost from the day you start flying school. We can go further in fifteen minutes than those people could in those days [1930s] with a full load of fuel." The general notes further that the old ground checkpoint method of navigation—following the Southern Pacific railroad, for example—is now totally inadequate. At present speeds and altitudes—sometimes above 40,000 feet—electronic navigation systems are necessary to aid air force pilots "to keep track of where we are, and help us get from one point to the next point" (interview, November 10, 1981).

Electronic surveillance further negates the freedom Thad Foster once enjoyed; picking up girl friends at some remote point and flying them to another is no longer practiced. The Federal Aviation Administration

(FAA) controls all aircraft movement above 1,800 feet and monitors on radar all flights below that level. "You never go any place that someone is not keeping track," General Nelson added. "In our command post we keep track of every airplane that takes off. . . . We know where they are and what they are doing one hundred percent of the time" (ibid.).

The current worldwide deployment of the air force created by global instability further eliminates many of the personal freedoms enjoyed in the pre–World War II Air Corps. And, in addition, escalating growth and expansion has brought with it escalating controls, checks, and balances—the General Accounting Office and the Office of Management and Budget, for example—to prevent waste, fraud, and abuse. And in the area of abuse, the mention of flying young ladies around in military aircraft brought a grimace to the general's usually placid face. "All you need to do," he explained, "is get your girl friend out toward an airplane, and somebody would write Jack Anderson and you'd make the headlines. Those kinds of personal freedoms are long gone" (ibid.).

Airports, like airplanes, have escalated in cost, and the public's response to their location has altered dramatically in the interim since Lt. John W. Egan helped dedicate the Baton Rouge, Louisiana, airport. Public sentiment frequently outweighs urban pride as expanding landing facilities are moved further and further from population centers. The public's objection to jet aircraft noise is articulated in the vernacular of another jet age phenomenon, the environmental impact study. In addition, airport dedications no longer attract the throngs that welcomed Lieutenant Egan at Baton Rouge. And if military aircraft are invited to provide aerial demonstrations, the request is carefully considered and most likely declined. The reason—cost. Should a unit find it feasible to participate in a fly-over at some public event, it is usually scheduled in conjunction with a training mission. "The dollars and cents of putting these airplanes in the sky" eliminates most requests of this nature, explained Lt. Col. Forrest S. Winebarger, commander of the 924th Tactical Fighting Group, USAF Reserve. If the aircraft under his command participate in a fly-over, "I have to get some training [benefits] out of these air crew members and out of these aircraft before I can support it" (interview, November 5, 1981).

A structure, of necessity, must be adapted to its function. Consequently, with the air force's advanced technology and material sophistication came more restrictive managerial procedures and the consequent decline of the once-cherished personal freedoms. The old-time officers of the Army Air Corps, wont to dream of the past, lament the passing of that golden era of military aviation. Maj. Gen. William L. Kennedy explained:

When I finished flying in 1961, I was so pleased that my love affair with airplanes was over, and the principal reason it was over was because of red tape. Sometimes you would spend more time making up a [flight] clearance than you would in making a trip. All the way [to your destination] you had to tell people where you are, when you expected to arrive at your destination, and all of that.

Then, pausing for a moment's reflection, he concluded, "But it was necessary" (interview, January 19, 1978).

Both time and technology have removed the present vast distances from the era of Johnson's Ranch. No place remains in the scheme of things for that brief moment in time when airplanes frequented the riverside airfield. From his adobe command post overlooking the Rio Grande, Elmo Johnson, for all practical purposes, commanded his own personal air force that included some of the Air Corps' finest pilots and the most modern aircraft in the world. He provided the landing field and living quarters for the crews, maintained the facilities, built the radio station, and was accredited to the Air Corps as a border watch. And while he never solicited his philosophical position as base commander, he never abused his rank. If a visiting pilot flew him up to Study Butte to talk business with a friend, the pilot could enter the flying time in his log book. The same was true if they flew him on sorties into Mexico to look for lost cattle and chase law breakers. (The latter escapade, of course, was never officially reported.) One former Air Corps pilot who frequented the Big Bend airfield explained the phenomenon of Johnson's Ranch in its simplest terms. Elmo, like his uniformed friends, "just liked to play airplane."

The era of Johnson's Ranch has come and gone, bringing with it changing concepts of man's role in military aviation and leaving in its wake a vastly different world in which manned flight dominates every aspect of life on this planet. In military, business, science, industry, and personal transportation, the airplane has emerged as essential. Even in the vast stretches of the Big Bend country where man's equestrian prowess once dominated the cattle kingdom, the airplane has again exacted its tribute. In December 1946, the Rawls family, Big Bend pioneer ranchers, held open house for their flying friends and their families during the holidays. A latter-day air force pilot flying over the Rawls' private landing strip and hangar commented in astonishment, "Seven planes and not a horse in sight!"[2]

---

2. Undated and unidentified newspaper clipping. Marfa, Texas, population 2,495, located in the heart of the West Texas cattle industry, claims more licensed pilots and private aircraft per capita than any municipality in the United States (*Austin American-Statesman*, November 16, 1980).

Yet some things remain static within this changing society. As the great river flows quietly past the crumbling ruins of Johnson's trading post, the sound of gunfire still punctures the pristine silence of that primitive land. The specter of drama, the tragedy of man versus the law, still persists along that remote section of the international border. Dr. Ross A. Maxwell explained:

> Smuggling was a way of life down there. If you get in the way of 'em, they'll kill you. If you don't bother 'em, most of 'em won't bother you.
>
> The Border Patrol has always kept a lookout along the river. In the twenties and thirties, it was cattle and booze. Now it's marijuana, guns, and ammunition. They can't stop it. There will always be some going across if the reward equals the risk. And when the rewards are high, a few men are going to die. (Interview, September 6, 1978).

Time touches everything and everybody. Panoramas blur with the passing decades as new vistas appear in their place. Airplanes no longer land at Johnson's Ranch. The river is silent; the era of organized banditry is long past, its drama relegated to history and to those who remember. And the intrepid airmen who once helped secure that volatile section of the international border have long since retired from the service.

Johnson's Ranch emerges in retrospect as part of an almost mythical existence that was destined to end; a transitory experience, just a brief place in time that came only once to the all too few who visited there. Like the ghostly mists that rise from the Rio Grande on autumn mornings and silently disappear somewhere amid the trees, and the mountains, and the canyons, so must this ranchland airfield. It brought security to the region and pleasure to the faithful, but somehow no longer fit into the ongoing scheme of things. It would ultimately live on, but only in the memories of the few who were privileged to have passed that way.

Bibliography

PRIMARY SOURCES

Correspondence

Alvarado, C. J. Presidio, Tex., July 8, 1980.
Andrews, Rayma L. El Paso, Tex., April 22, 1981.
Aplington, Col. Henry, II (USMC, Ret.). Warner, N.H., January 9, 1980.
Ashcroft, Lt. Col. Lucas J. (USAF, Ret.). San Antonio, Tex., December 9, 1979.
Ashley, Thelma. San Antonio, Tex., August 1, 1980.
Babb, Annie Lee. Sanderson, Tex., January 21, 1979, undated.
Barlow, Riley M. Marfa, Tex., August 29, 1980, November 24, 1980, February 26, 1981.
Crone, Helen. County and district clerk, Brewster County, Alpine, Tex., January 9, 1978, December 29, 1978, March 18, 1980.
Eastman, James N., Jr. Chief, Research Division, Albert F. Simpson Historical Research Center, Maxwell Air Force Base, Ala., October 23, 1980, December 16, 1980.
Finnell Scott, Patricia. Archives of the Big Bend, Sul Ross State University, Alpine, Tex., October 10, 1978, October 18, 1978, February 15, 1979, March 8, 1979, August 8, 1979, August 30, 1979, September 13, 1979.
Freyer, Lt. Col. Fred R. (USAF, Ret.). St. Simons Island, Ga., September 20, 1978, October 10, 1978, January 5, 1979, January 17, 1979, November 11, 1980.
Green, Aaron A. Alpine, Tex., November 5, 1978.
Green, Lorene. Alpine, Tex., August 23, 1980, August 31, 1980, September 30, 1980, November 3, 1980, November 14, 1980.
Hammack, Lt. Col. Robert (CAP). El Paso, Tex., August 4, 1980.
Hansen, Col. George W. (USAF, Ret.). Corvallis, Oreg., September 18, 1978.
Harrison, George W. Port Isabel, Tex., September 26, 1980.
Hicks, Maj. Gen. Joseph H. (USAF, Ret.). Medford, Oreg., December 15, 1979, January 2, 1980, January 7, 1980, February 10, 1980.

Hopper, Col. L. E. (CAR, Ret.). New Orleans, La., extensive correspondence, 1983.

Hudson, Col. LeRoy (USAF, Ret.). Leucadia, Calif., September 20, 1978, September 29, 1978, December 16, 1978, January 2, 1979, September 1979.

Johnson, Maj. Gen. Harry W. (USA, Ret.). Travelers Rest, S.C., March 2, 1980.

Johnson, Maj. Kyle (USAF, Ret.). Riverside, Calif., June 19, 1979.

Kasper, John F. Corpus Christi, Tex., January 13, 1978, September 15, 1978, August 28, 1979, November 27, 1980.

Kelley, Lois Neville. El Paso, Tex., October 24, 1980, January 26, 1981, February 26, 1981, March 12, 1981, March 31, 1981, April 17, 1981.

Macrum, Brig. Gen. Robert S. (USAF, Ret.). Acton, Maine, October 8, 1978, March 2, 1979.

Madrid, Lucia Rede. Redford, Tex., April 30, 1980.

Mahoney, Tom. Poughkeepsie, N.Y., October 15, 1980, November 9, 1980, November 28, 1980, December 16, 1980, May 5, 1981.

Morton, Glenn B. Glendale, Calif., May 18, 1981.

Mueller, William. El Paso, Tex., June 28, 1980, July 17, 1980.

Myers, Lt. Gen. Samuel L. (USA, Ret.). Del Rio, Tex., January 14, 1980, January 31, 1980, March 19, 1980, March 28, 1980, August 14, 1980.

Newman, Bud. The University of Texas at El Paso, March 14, 1980, January 7, 1981.

Phillips, Col. Henry L. (USAF, Ret.). San Antonio, Tex., June 19, 1979.

Rhine, Bertrand. Pasadena, Calif., July 15, 1980, August 5, 1980, December 8, 1980.

Rinehart, Lt. Col. Howard E. (USAF, Ret.). Arlington, Va., April 10, 1978, November 10, 1980, June 16, 1981.

Robey, Maj. Gen. P. H. (USAF, Ret.). Tucson, Ariz., January 24, 1978, November 14, 1978, November 15, 1980.

Rodieck, Col. George E. (USAF, Ret.). Green Valley, Ariz., February 12, 1979, April 19, 1979.

Rose, Col. Elmer P. (USAF, Ret.). San Antonio, Tex., February 5, 1979, April 30, 1979.

Smith, David L. Corpus Christi, Tex., December 7, 1978, December 11, 1978, December 18, 1978, January 11, 1979, June 28, 1979, July 23, 1980, August 19, 1980, November 18, 1980, December 1, 1980, December 17, 1980, December 22, 1980.

Smith, Lt. Col. Samuel G. (USA, Ret.). San Antonio, Tex., December 9, 1979, April 14, 1980.

Smithers, W. D. El Paso, Tex., seventeen dated and one undated communications between June 22, 1975 and September 7, 1979.

Stillwell, Hallie. Alpine, Tex., October 10, 1978, November 6, 1978.

Swift, Roy L. Floresville, Tex., January 20, 1979.

Interviews

Andrews, Rayma L. El Paso, Tex., February 18, 1981, July 2, 1981, September 24, 1981.

————. Transcript of oral interview conducted by Maj. L. W. Voit in the Oral History Collection, James Gillam Gee Library, East Texas State University, Commerce, Tex., December 29, 1977.

Ashcroft, Lt. Col. Lucas J. (USAF, Ret.). San Antonio, Tex., January 29, 1980.

Ashley, E. G. San Antonio, Tex., January 28, 1980, December 5, 1980, January 2, 1982.

Ashley, Thelma. San Antonio, Tex., January 10, 1979, January 28, 1980.

Babb, G. E. Sanderson, Tex., June 3, 1966.

Cartledge, Robert L. Austin, Tex., May 7, 1965.

Cartledge, Wayne. Marfa, Tex., December 29, 1964.

Casner, Katheryn. Austin, Tex., January 16, 1978, March 30, 1978.

Deerwester, Col. Charles W. (USAF, Ret.). San Antonio, Tex., February 11, 1971.

Dichter, Mrs. Juanita. Seaside, Oreg., February 19, 1984.

Egan, Col. John W. (USAF, Ret.). Bandera Falls, Tex., February 7, 1978, February 20, 1979, April 30, 1980.

Green, Aaron A. Alpine, Tex., September 10, 1979.

Harrison, George W. Port Isabel, Tex., November 26, 1980.

Hubbell, Col. Frank (USAF, Ret.). Topeka, Kans., April 13, 1980.

Johnson, Elmo. Sonora, Tex., June 3, 1966.

Johnson, Maj. Kyle (USAF, Ret.). Riverside, Calif., September 8, 1979.

Johnson, Robert. November 1980, undated.

Kasper, John F. Corpus Christi, Tex., September 24, 1979.

Kennedy, Maj. Gen. William L. (USAF, Ret.), Bandera Falls and San Antonio, Tex., January 19, 1978, January 7, 1979, February 20, 1979, September 11, 1979, January 9, 1981, July 30, 1981.

Macrum, Brig. Gen. Robert S. (USAF, Ret.). Acton, Maine, five undated audio tapes, 1979, 1980.

Maxwell, Ross A. Austin, Tex., January 30, 1978, September 6, 1978, November 29, 1978.

Mayhue, Col. Don W. (USAF, Ret.). San Antonio, Tex., February 7, 1978.

Morton, Glenn B. Glendale, Calif., June 1, 1981.

Myers, Lt. Gen. Samuel L. (USA, Ret.). Del Rio, Tex., Seven undated audio tapes, 1980, 1981.

Nelson, Lt. Gen. William R. (USAF). Bergstrom Air Force Base, Tex., November 10, 1981.

Nolta, Mrs. Dale E. Willows, Calif., May 29, 1981.

Parker, Maj. Gen. Hugh A. (USAF, Ret.). San Antonio, Tex., February 8, 1978, April 13, 1978, September 20, 1979, May 12, 1980.

Phillips, Col. Henry L. (USAF, Ret.). San Antonio, Tex., January 29, 1980.

Pitts, Joel G. Dallas, Tex., November 16, 1980.

Rhine, Bertrand. Pasadena, Calif., May 13, 1980.

Rinehart, Lt. Col. Howard E. (USAF, Ret.). Arlington, Va., September 10, 1980.
Robey, Maj. Gen. P. H. (USAF, Ret.). Tucson, Ariz., February 16, 1981.
Rose, Col. Elmer P. (USAF, Ret.). San Antonio, Tex., December 10, 1979, January 28, 1980.
Sigerfoos, Col. Edward (USAF, Ret.). San Antonio, Tex., February 8, 1978.
Skiles, Jack, Sr., to Mavis Bryant. Langtry, Tex., 1977. Tapes 20 and 21, Natural Areas Survey, Texas Conservation Foundation, Austin, Tex.
Smith, Lt. Col. Samuel G. (USA, Ret.). San Antonio, Tex., January 30, 1980, January 22, 1981.
Smithers, W. D. El Paso, Tex., August 27, 1975.
Tanger, Leigh. Pineville, Ark., April 15, 1980, November 29, 1980.
Ward, Col. Roy P. (USAF, Ret.). April 9, 1980, May 3, 1981.
Wier, J. Rex, Jr. Austin, Tex., April 9, 1980.
Winebarger, Lt. Col. Forrest S. (USAF). Bergstrom Air Force Base, Tex., November 5, 1981.
Yarborough, Ralph W. Austin, Tex., November 25, 1980.

Manuscript Collections and Public Records

Hopper, Col. L. E., chairman, CAP National Historical Committee. Personal collection. New Orleans, La.
Myers, Lt. Gen. Samuel L. Papers. Del Rio, Tex.
National Archives. Civil Aeronautics Administration, Project Application File, 1935–1938, for Johnson's Ranch, Tex. Record Group 237.
———. Quartermaster's Correspondence Files. Record Group 92.
———. Records of the Adjutant General's Office, 1780s–1917. Record Group 94.
———. Records of the Adjutant General's Office, 1917–   . Record Group 407.
———. Records of the Army Air Force, Central Document File, 1917–1938, Project Files, Corps Area, Eighth Corps, 601.53 Leases, Box 2983. Record Group 18.
———. Records of the Department of State Pertaining to the Internal Affairs of Central America, 1910–1929. Record Group 59.
———. Records of the Southern Defense Command, 1941–1944. Record Group 338.
———. Records of the United States Army Continental Commands, 1920–1942, Camp Marfa. Record Group 394.
———. United States Department of Labor, Immigration and Naturalization Service. Record Group 85.
Office of the County and District Clerk. Deed Records. Brewster County, Alpine, Tex.
———. Land Records. Brewster County, Alpine, Tex.
Smithers, W. D. Manuscript collection. Humanities Research Center, University of Texas at Austin.
Sterling, Governor Ross S. Papers. Texas State Archives, Austin, Tex.

Texas Parks and Wildlife Department, Records Center. Record of Land Acquisitions, Big Bend National Park. Austin, Tex.

SECONDARY SOURCES

Books, Articles, Documents, and Theses

Aikman, Duncan. "50,000 Cannon Balls." *Harper's Monthly Magazine* 159 (July 1929): 243–250.
———. "A General's War." *The Outlook and Independent*, March 27, 1929, p. 488.
Bilstein, Roger. *Flight Patterns: Trends of Aeronautical Development in the United States, 1918–1929.* University of Georgia Press, 1984.
Boyne, Walt. "Fly-Off!" *Airpower* 9 (September 1979): 32–45.
Braddy, Haldeen. *Pershing's Mission in Mexico.* El Paso: Texas Western Press, 1966.
"Brighter Skies beyond the Border." *The Literary Digest*, May 18, 1929, p. 10.
Casey, Clifford B. *Mirages, Mysteries, and Reality—Brewster County, Texas—The Big Bend of the Rio Grande.* Hereford, Tex.: Pioneer Book Publishers, 1972.
Cook, Mary Katherine. "W. D. Smithers, Photographer-Journalist." Master's thesis, University of Texas at Austin, 1975.
Cox, Mike. *Red Rooster Country.* Hereford, Tex.: Pioneer Book Publishers, 1970.
Cullinane, Daniel S. "Ciudad Juárez and the Escobar Revolution." *Password* 3 (July 1958): 97–106.
Daniels, Josephus. *Shirt-Sleeve Diplomat.* Chapel Hill: University of North Carolina Press, 1947.
Davis, R. E. G. *Airlines of the United States since 1914.* London: Putnam, 1972.
Dean, Jack. "The Secret of the Shrike." *Airpower* 9 (September 1979): 8–17.
DeConde, Alexander. *Herbert Hoover's Latin-American Policy.* Stanford, Calif.: Stanford University Press, 1951.
Densford, M. Sgt., ed. *From Jennies to Jets: The Story of the 111th Squadron.* National Guard Association of Texas, 1973.
Diccionario Porrúa de historia, biografía, y geografía de México. Vol. 1. Mexico, 1976.
Dulles, John W. F. *Yesterday in Mexico: A Chronicle of the Revolution, 1919–1936.* Austin: University of Texas Press, 1961.
Earney, Ann. "Bears, Beeves, and Bibles." *Texas Historian* 37 (January 1977): 1–7.
Farley, Kevin. "Terrell County Airfields." *Texas Historian* 31 (May 1971): 21–23.
Fierro Villalobos, Roberto. *Este es mi vida.* Mexico, 1964.
"The First Regiment of Cavalry." *Voice of the Mexican Border* (September 1933): 55–60.
Futrell, Robert F. *Command of Observation Aviation: A Study in Control of Tac-*

tical Airpower. USAF Historical Division Study. Maxwell Air Force Base, Ala.: Air University, 1956.

Glines, Lt. Col. Carroll V. *The Compact History of the United States Air Force.* New York: Hawthorn Books, 1963.

———. *The Modern United States Air Force.* Princeton, N.J.: D. Van Nostrand Co., 1963.

González y González, Luis. *Fuentes de la historia contemporánea de México: Libros y folletos.* 3 vols. Mexico City: El Colegio de México, 1961–1962.

Gray, C. G., ed. *Jane's All the World's Aircraft of 1929.* London: Sampson Low, Marston and Co., Ltd., 1929.

Gruening, Ernest. "The Recurring Rebellion in Mexico." *The New Republic,* March 27, 1929, pp. 161–165.

Hacket, Charles W. "Collapse of the Mexican Rebellion." *Current History* (June 1929): 499.

———. "The Mexican Rebellion." *Current History* (April 1929): 140.

Hinkle, Stacy C. *Wings and Saddles: The Air and Cavalry Punitive Expedition of 1919.* Southwestern Studies, Monograph 19. El Paso: Texas Western Press, 1967.

———. *Wings over the Border: The Army Air Service Armed Patrol of the United States–Mexico Border, 1919–1921.* Southwestern Studies, Monograph 26. El Paso: Texas Western Press, 1970.

Hopkins, George D. *The Airline Pilots: A Study in Elite Unionization.* Cambridge, Mass.: Harvard University Press, 1971.

"How General Escobar Fought, Ran Away, and Lived to Run Away Some Other Day." *Every Week Magazine,* February 14–15, 1931, p. 2.

Jablonski, Edward. *Flying Fortress.* Garden City, N.Y.: Doubleday, 1965.

LaMar, Nelson. "Flying the Tunnel 'Underground' in the Grand Canyon." *The Journal of Arizona History* 14 (Winter 1974): 319–324.

Lieuwen, Edwin. *Mexican Militarism: The Political Rise and Fall of the Revolutionary Army.* Albuquerque: University of New Mexico Press, 1968.

Lindbergh, Charles A. *We.* New York: G. P. Putnam's Sons, 1927.

Madison, Virginia. *The Big Bend Country of Texas.* Rev. ed. New York: October House, 1968.

Mahoney, Tom. "Into the Clouds for a Lost Cause." *Air Travel News,* December 1929, pp. 13–15.

———. "Reporters to the Rebels." *The Quill* (May 1930): 10–12.

Maurer Maurer, ed. *Air Force Combat Units of World War II.* Air University, Department of the Air Force, Washington, D.C.: U.S. Government Printing Office, 1961.

Maxwell, Ross A. *The Big Bend of the Rio Grande: A Guide to the Rocks, Landscape, Geologic History, and Settlers of the Area of the Big Bend National Park.* Austin: University of Texas Bureau of Economic Geology, Guidebook 7, 1968.

"Mexico's Airway to Harmony." *The Literary Digest,* March 15, 1930, p. 22.

Miller, Jay. "Gergor's Geldings." *Aerophile* 2, no. 1 (June 1979): 30.

Mohun, Philip. "The Revolution Racket." *Liberty,* July 21, 1934, pp. 32–36.

Moran, Gerard P. *Aeroplanes Vought.* Temple City, Calif.: Historical Aviation Album, 1978.

Mumme, Stephen P. "The Battle of Naco: Factionalism and the Conflict in Sonora, 1914–1915." *Arizona and the West* 21 (Summer 1979): 157–186.
Myers, Lt. Gen. Samuel L. "Oh! How the Horses Laughed!" *Armor* 85 (July–August 1976): 30–34.
Neprud, Robert E. *Flying Minute Men*. Duel, Sloan and Pearce, 1948.
Newton, Wesley Phillips. *The Perilous Sky: United States Aviation Diplomacy and Latin America, 1919–1931*. Coral Gables: University of Miami Press, 1978.
———. "The Role of Aviation in Mexican–United States Relations, 1912–1929." In Eugene R. Huck and Edward H. Moseley, eds., *Militarists, Merchants, and Missionaries: United States Expansion in Middle America*. University: University of Alabama Press, 1970.
Oakes, Claude M., comp. *Aircraft of the National Air and Space Museum*. Washington, D.C.: Smithsonian Institution Press, 1981.
Obregón, Alvaro. *Ocho mil kilómetros en campaña*. Mexico City: Librería en la Vad. de Ch. Bouret, 1917.
Papers Relating to the Foreign Relations of the United States, 1929. 3 vols. 71st Cong., 1st sess., H. Doc. 517. Washington, D.C.: U.S. Government Printing Office, 1943.
Parkes, Henry Bamford. *A History of Mexico*. Boston: Houghton Mifflin Co., 1938.
Ragsdale, Kenneth Baxter. *Quicksilver: Terlingua and the Chisos Mining Company*. College Station: Texas A & M University Press, 1976.
"A Ringside Seat for Mexican Revolutions." *The Literary Digest*, April 13, 1929, pp. 46–48.
Ross, Stanley R. "Dwight Morrow and the Mexican Revolution." *Hispanic American Review* 38 (November 1958): 506–528.
———. *Fuentes de la historia contemporánea de México: Periódicos y revistas*. 2 vols. Mexico City: El Colegio de México, 1965–1967.
Seagrave, Sterling, ed. *Soldiers of Fortune*. Time-Life Epic of Flight Series. Alexandria, Va.: Time-Life Books, 1981.
Setzler, Frank M. "Prehistoric Cave Dwellers in Texas." In *Explorations and Field-Work of the Smithsonian Institution in 1932*. Washington, D.C.: Smithsonian Institution, 1933.
Sherrod, Robert. *History of Marine Corps Aviation in World War II*. Washington, D.C.: Combat Forces Press, 1952.
Sims, Edward H. *The Aces Talk*. New York: Ballantine Books, 1974.
Smithers, W. D. "Bandit Raids in the Big Bend Country." *Sul Ross State College Bulletin* 43 (September 1963): 74–105.
———. "The Border Trading Posts." *Sul Ross State College Bulletin* 41 (September 1961): 41–56.
———. *Chronicles of the Big Bend*. Austin: Madrona Press, 1976.
Stirling, W. M. *Explorations and Field-Work of the Smithsonian Institution in 1930*. Publication 3111. Washington, D.C.: Smithsonian Institution, 1931.
Swanborough, F. G. *United States Military Aircraft since 1909*. New York: Putnam, 1963.
Tannenbaum, Frank. *Peace by Revolution*. New York: Columbia University

Press, 1933.

This Fabulous Century, vol. IV, 1930–1940. Ezra Bowen, series ed. New York: Time-Life Books, 1969.

Tyler, Ronnie C. *The Big Bend: A History of the Last Texas Frontier.* Washington, D.C.: National Parks Service, 1975.

United States Civil Aeronautics Authority. *Directory of Airports and Seaplane Bases, Part IV.* Civil Aeronautics Bulletin no. 11. Washington, D.C.: U.S. Government Printing Office, 1939.

United States Department of the Treasury. *Documents Pertaining to Foreign Funds Control.* General Ruling no. 14, August 14, 1942. Washington, D.C.: U.S. Government Printing Office, 1944.

Valenzuela, René A. "Chihuahua, Calles and the Escobar Revolt of 1929." Master's thesis, University of Texas at El Paso, 1975.

Villela, José, Jr. *Pioneros de la aviación mexicana.* Mexico: Ediciones Colofón, 1964.

Wagner, Ray. *American Combat Planes.* Garden City, N.Y.: Hanover House, 1960.

Ward, John W. "The Meaning of Lindbergh's Flight." In Henning Cohen, ed., *The American Culture.* Boston: Houghton Mifflin, 1968.

Weigley, Russell F. *The American Way of War: A History of United States Military Strategy and Policy.* New York: Macmillan, 1973.

West, James E. *The Lone Scout of the Sky.* Philadelphia: John D. Winston Co., 1928.

"What the Mexicans Are Fighting Over." *The Literary Digest*, April 13, 1929, p. 11.

Newspapers and Journals

Aero Digest
Air Corps News Letter
Airpower
Alpine Avalanche
Austin American-Statesman
Big Bend Sentinel (Marfa, Tex.)
Buffalo (New York) Times
Cavalry Journal
Douglas (Arizona) Daily Dispatch
El Paso Herald

El Paso Herald-Post
El Paso Times
Laredo Times
The New Era (Marfa, Tex.)
New York Times
San Angelo Evening-Standard
San Angelo Standard-Times
San Antonio Express
San Antonio Light
Texas Business Review

Index